Collins

KS3
Maths
Foundation Level

Revision Guide

Samya Abdullah, Rebecca Evans
and Gillian Spragg

About this Revision & Practice book

When it comes to getting the best results, practice really does make perfect!

Experts have proved that repeatedly testing yourself on a topic is far more effective than re-reading information over and over again. And, to be as effective as possible, you should space out the practice test sessions over time.

This Complete Revision & Practice book is specially designed to support this approach to revision and includes seven different opportunities to test yourself on each topic, spaced out over time.

Revise

These pages provide a recap of everything you need to know for each topic.

You should read through all the information before taking the Quick Test at the end. This will test whether you can recall the key facts.

> **Quick Test**
>
> 1. Simplify $4x + 7y + 3x - 2y + 6$
> 2. Simplify $c \times c \times d \times d$
> 3. Find the value of $4x + 2y$ when $x = 2$ and $y = 3$

Practise

These topic-based questions appear shortly after the revision pages for each topic and will test whether you have understood the topic. If you get any of the questions wrong, make sure you read the correct answer carefully.

Review

These topic-based questions appear later in the book, allowing you to revisit the topic and test how well you have remembered the information. If you get any of the questions wrong, make sure you read the correct answer carefully.

Mix it Up

These pages feature a mix of questions for all the different topics, just like you would get in a test. They will make sure you can recall the relevant information to answer a question without being told which topic it relates to.

Test Yourself on the Go

Visit our website at **collins.co.uk/collinsks3revision** and print off a set of flashcards. These pocket-sized cards feature questions and answers so that you can test yourself on all the key facts anytime and anywhere. You will also find lots more information about the advantages of spaced practice and how to plan for it.

Workbook

This section features even more topic-based and mixed test-style questions, providing two further practice opportunities for each topic to guarantee the best results.

This revision guide and workbook are suitable for students working upto the expected level in KS3 Maths.

Symbols are used to highlight questions that test key skills:

(MR) Mathematical Reasoning   (PS) Problem Solving   (FS) Financial Skills

Try to answer as many questions as possible without using a calculator.

Questions where calculators **must not** be used are marked with this symbol:

Contents

			Revise	Practise	Review

N Number   **A** Algebra   **G** Geometry and Measures

S Statistics   **P** Probability   **R** Ratio, Proportion and Rates of change

Contents

N Number   **A** Algebra   **G** Geometry and Measures

Contents

S Statistics   **P** Probability   **R** Ratio, Proportion and Rates of change

Number 1

You must be able to:

- Order positive and negative numbers
- Use the symbols =, ≠, <, >, ≤, ≥
- Multiply and divide integers
- Carry out operations following BIDMAS.

Positive and Negative Numbers

- A **number line** can be used to order **integers**.

Example

Place 7, 5, –6, –1 and 3 in ascending order.

Going from left to right, you can see that the ascending order is –6, –1, 3, 5, 7.

- **Place value** can be used to compare the size of large numbers.

Example

Which is greater, 3408 or 3540?

Number	Thousands	Hundreds	Tens	Units
3408	3	4	0	8
3540	3	5	4	0

Both numbers have 3 thousands, but 3540 has 5 hundreds and 3408 only has 4 hundreds. Therefore 3540 is greater than 3408.

Key Point

Always compare digits from left to right.

- **Symbols** are used to state the relationship between two numbers.

Symbol	Meaning
>	Greater than
<	Less than
≥	Greater than or equal to
≤	Less than or equal to
=	Equal to
≠	Not equal to

Example

3540 is greater than 3408 can be written as 3540 > 3408.

Multiplication and Division

- To multiply big numbers using the grid method, partition both numbers into their hundreds, tens and units.

Example

Calculate 354×273

×	**300**	**50**	**4**
200	60 000	10 000	800
70	21 000	3500	280
3	900	150	12

> 354 is made up of 3 hundreds (300), 5 tens (50) and 4 units (4). 273 is made up of 2 hundreds (200), 7 tens (70) and 3 units (3).

> Complete the multiplication grid, for example $50 \times 70 = 3500$ and $4 \times 3 = 12$

Then add together the numbers in the grid:

$60\,000 + 10\,000 + 800 + 21\,000 + 3500 + 280 + 900 + 150 + 12$
$= 96\,642$

So $354 \times 273 = 96\,642$

- Division can also be broken down into steps.

Example

Calculate $762 \div 3$

$3\overline{)7\,6\,2}$	$3\overline{)7\,{}^16\,2}$ 2	$3\overline{)7\,{}^16\,{}^12}$ 25	$3\overline{)7\,{}^16\,{}^12}$ 254
Set up your division.	Work out how many 3s go into 7. 3 goes into 7 twice with 1 left over. You carry the 1 over.	Work out how many 3s go into 16. 3 goes into 16 five times with 1 left over. You carry the 1 over.	Work out how many 3s go into 12. 3 goes into 12 four times with zero left over. So $762 \div 3 = 254$

BIDMAS

- BIDMAS gives the order in which operations should be carried out:

Brackets
Indices
Divide
Multiply
Add
Subtract

Example

$2 \times 6 + 5 \times 4 = 32$

Key Point

Indices are also called powers.

Quick Test

1. Put 3750, 3753, 3601, 3654 and 3813 in ascending order.
2. Work out 435×521
3. Work out $652 \div 4$
4. Work out $4 \times 3 + 7 \times 4$
5. Write the following using a mathematical symbol: -5 is less than 3.

Key Words

integer
positive
negative
place value

Number 2

You must be able to:

- Understand square numbers and square roots
- Write a number as a product of prime factors
- Find the lowest common multiple and highest common factor.

Squares and Square Roots

- **Square numbers** are calculated by multiplying a number by itself.

> **Example**
>
> $5^2 = 5 \times 5 = 25$
>
> The first ten square numbers are:
> 1, 4, 9, 16, 25, 36, 49, 64, 81 and 100.

- A **square root** ($\sqrt{}$) is the inverse or opposite of a square.

> **Example**
>
> $\sqrt{36} = 6$

- There are other powers as well as squares.

> **Example**
>
> $5^3 = 5 \times 5 \times 5 = 125$
>
> $5^4 = 5 \times 5 \times 5 \times 5 = 625$
>
> $5^5 = 5 \times 5 \times 5 \times 5 \times 5 = 3125$

Prime Factors

- **Factors** are numbers you can multiply together to make another number.
- Every number can be written as a **product** of prime factors.
- A **prime** number has exactly two factors, itself and 1.

> **Example**
>
> 45 can be expressed as a product of prime factors.
>
> This can be done using a **prime factor tree**:
>
>
>
> $45 = 5 \times 9$
> and
> $9 = 3 \times 3$
>
> So $45 = 3 \times 3 \times 5$

Key Point

Product means multiply.

Key Point

A prime factor tree breaks a number down into its prime factors.

Always start by finding a prime number which is a factor, in this case 5.

Remember to write the final answer as a product.

Lowest Common Multiple and Highest Common Factor

- The **lowest common multiple** (LCM) is the lowest multiple two or more numbers have in common.
- The **highest common factor** (HCF) is the highest factor two or more numbers have in common.

Example

Find the lowest common multiple and highest common factor of 12 and 42.

Write both numbers as a product of prime factors.

$12 = 2 \times 2 \times 3$ $42 = 2 \times 3 \times 7$

Complete the Venn diagram.

12   **42**

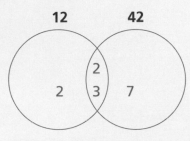

Common factors are placed in the overlap.

The LCM is the product of all the numbers in both circles.

LCM $= 2 \times 2 \times 3 \times 7 = 84$

The HCF is the product of the numbers in the overlap.

HCF $= 2 \times 3 = 6$

- LCM and HCF are used to solve many everyday problems.

Example

Thomas is training to swim the English Channel. He has to visit his doctor every 12 days and his nutritionist every 15 days. If on 1st October he has both appointments on the same day, on what date will he next have both appointments on the same day?

Find the LCM of 12 and 15 – this is 60.

60 days after 1st October is 30th November.

Key Words

square number
square root
factor
product
prime
lowest common multiple
highest common factor

Quick Test

1. Write down the value of 7^2.
2. Write down the value of $\sqrt{49}$.
3. Write 40 as a product of prime factors.
4. Find the lowest common multiple of 14 and 36.
5. Find the highest common factor of 24 and 32.

Algebra

Sequences 1

You must be able to:

- Use a function machine to generate terms of a sequence
- Recognise arithmetic sequences
- Generate sequences from a term to term rule.

Function Machines

- A **function machine** takes an input, applies one or more operations and produces an output.

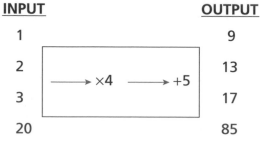

INPUT	OUTPUT
1	9
2	13
3	17
20	85

Sequences

- A **sequence** is a set of shapes or numbers which follow a pattern or rule.
- The outputs from a **function machine** form a sequence.

Example

In the sequence below, the next pattern is formed by adding an extra layer of tiles around the previous pattern.

| Basic design | Layer 1
6 new tiles | Layer 2
10 new tiles | Layer 3
14 new tiles |

With each new layer the number of new tiles needed increases by 4. This pattern can be used to predict how many tiles will be needed to make larger designs.

- An **arithmetic sequence** is a set of numbers with a common difference between consecutive terms.

Example

4, 7, 10, 13, 16,… is an arithmetic sequence with a common difference of 3.

12, 8, 4, 0, –4,… is also an arithmetic sequence. It has a common difference of –4.

Key Point

Arithmetic sequences are used to solve many real-life problems.

- In a **geometric sequence**, each term is multiplied by the same value to find the next term.

> **Example**
> 2, 4, 8, 16, 32, 64,… is a geometric sequence where each term is multiplied by 2 to find the next term.

- There are many other sequences of numbers that follow a pattern.

> **Example**
> 1, 4, 9, 16, 25, 36,… are square numbers.
>
> 1, 3, 6, 10, 15, 21,… are known as the triangular numbers.
>
> 1, 1, 2, 3, 5, 8, 13,… is known as the Fibonacci sequence. ← Each number is the sum of the previous two numbers.

Finding Missing Terms

- The **term to term rule** links each term in the sequence to the previous term.

> **Example**
> 5, 8, 11, 14, 17,…
>
> In this set of numbers the next term is found by adding 3 to the previous term. Therefore the term to term rule is +3.
>
> This rule can be used to find the next numbers in the sequence.
>
> $17 + 3 = 20$ $20 + 3 = 23$ $23 + 3 = 26$
>
> Therefore the next three terms in the sequence are 20, 23, 26,…

- The term to term rule can also be used to find missing terms.

> **Example**
> 13, 17, 21, _____, 29,…
>
> The term to term rule is +4 and so the missing term is 25.

> **Quick Test**
>
> 1. Find the missing inputs and outputs in the function machine.
>
>   INPUT   OUTPUT
>
>   6   ☐
>
>   ☐ ⟶ ÷2 ⟶ +1   25
>
> 2. Write down the next five terms in this sequence:
>   5, 9, 13, 17,…
> 3. Write down the term to term rule in this sequence.
>   24, 12, 6, 3, 1.5,…
> 4. Which of the following sequences is arithmetic?
>   A) 4, 6, 9, 13, 18,…   B) 3, 6, 9, 12, 15,…
> 5. Find the missing term in the following sequence of numbers.
>   21, 27, 33, _____, 45, 51,…

> **Key Words**
>
> function machine
> sequence
> arithmetic sequence
> geometric sequence
> term to term

Sequences 2

You must be able to:

- Generate terms of a sequence from a position to term rule
- Find the nth term of an arithmetic sequence
- Recognise quadratic sequences.

The nth Term

- The nth term is also called the **position to term rule**.
- It is an algebraic expression that represents the operations carried out by a function machine.

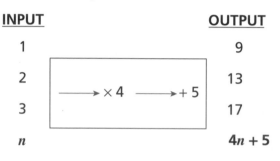

INPUT		OUTPUT
1		9
2	$\times 4 \longrightarrow + 5$	13
3		17
n		$4n + 5$

- The **nth term** can be used to generate terms of a sequence.

Example

The nth term of a sequence is given by $3n + 5$.

To find the first term you **substitute** $n = 1$

$3 \times 1 + 5 = 8$ 8 is the first term in the sequence.

To find other terms, you can substitute different values of n.

When $n = 2$	When $n = 3$	When $n = 4$
$3 \times 2 + 5 = 11$	$3 \times 3 + 5 = 14$	$3 \times 4 + 5 = 17$
So 11 is the second term in the sequence.	So 14 is the third term in the sequence.	So 17 is the fourth term in the sequence.

The nth term $3n + 5$ produces the sequence of numbers:

8, 11, 14, 17, 20,…

The rule can be used to find any term in the sequence. For example, to find the 50th term in the sequence substitute $n = 50$

$3 \times 50 + 5 = 155$

Key Point

For the first term in the sequence, n always equals 1.

Finding the nth Term

- To find the nth term, look for a pattern in the sequence of numbers.

> **Example**
> The first five terms of a sequence are 7, 11, 15, 19, 23.
>
> **The term to term rule** is +4 so the nth term starts with $4n$.
>
> The difference between $4n$ and the output in each case is 3, so the final rule is $4n + 3$.
>
> | Input | × 4 | Output |
> |-------|-----|--------|
> | 1 | 4 | 7 |
> | 2 | 8 | 11 |
> | 3 | 12 | 15 |
> | 4 | 16 | 19 |
> | 5 | 20 | 23 |
> | n | $4n$ | $4n + 3$ |

Quadratic Sequences

- **Quadratic** sequences are based on square numbers.

> **Example**
> The first five terms of the sequence $2n^2 + 1$ are as follows:
>
> When $n = 1$ $2 \times (1)^2 + 1 = 3$
> When $n = 2$ $2 \times (2)^2 + 1 = 9$
> When $n = 3$ $2 \times (3)^2 + 1 = 19$
> When $n = 4$ $2 \times (4)^2 + 1 = 33$
> When $n = 5$ $2 \times (5)^2 + 1 = 51$
>
> The sequence starts 3, 9, 19, 33, 51,…

- Triangular numbers are produced from a quadratic sequence.

> **Key Point**
>
> Use BIDMAS when calculating terms in a sequence.

> **Quick Test**
>
> 1. Write down the first five terms in the sequence $5n + 3$.
> 2. Write down the first five terms in the sequence $n^2 + 4$.
> 3. a) Find the nth term for the following sequence of numbers:
>   6, 9, 12, 15, 18,…
>   b) Find the 50th term in this sequence.
> 4. What is the nth term also known as?

> **Key Words**
>
> nth term
> position to term
> substitute
> quadratic

 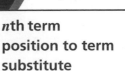

Review Questions

Key Stage 2: Key Concepts

1 Which is closer to 2000?

1996 or 2007

Explain how you know. [2]

2 Calculate 476 − 231 [calculator icon] [2]

3 Complete the table below by rounding each number to the nearest 1000.

	To nearest 1000
4587	
45 698	
457 658	
45 669	

[2]

4 Write these in order starting with the smallest.

0.56   55% $\frac{27}{50}$ 0.6   0.63 [3]

5 Ahmed is twice as old as Rebecca.

Rebecca is three years younger than John.

John is 25 years old.

How old is Ahmed? [2]

6 Calculate 467 × 34 [calculator icon] [2]

7 Calculate 156 ÷ 3 [calculator icon] [2]

8 On the scale below draw arrows to show 1.6 and 3.8

[scale marked from 1 to 4]

[2]

Total Marks _____ / 17

1 A bottle holds 1 litre of fizzy drink. Mariam pours four glasses for her friends.

Each glass contains 200ml.

How much fizzy drink is left in the bottle?   [3]

2 Below are five digit cards.

| 7 | 5 | 1 | 6 | 3 |

Choose two cards to make the following two-digit numbers.

   a) A square number   [1]

   b) A prime number   [1]

   c) A multiple of 6   [1]

   d) A factor of 60   [1]

3 An equilateral triangle has a perimeter of 27cm.

What is the length of one of its sides?   [2]

4 Two-thirds of a number is 22.

What is the number?   [2]

5 Here is an isosceles triangle drawn inside a rectangle.

Find the value of the angle x.   [3]

6 S and T are two whole numbers.

$S + T = 500$

S is 100 greater than T.

Find the value of S and T.   [2]

Total Marks _____ / 16

Practice Questions

Number

1 The table shows the average daily minimum temperature and the average daily maximum temperature in a town for every month of the year (given to the nearest °C). 📱

Month	Jan	Feb	Mar	Apr	May	Jun	Jul	Aug	Sep	Oct	Nov	Dec
Average Daily Minimum Temp.	−8	−4	−1	3	8	11	14	13	9	3	−2	−6
Average Daily Maximum Temp.	−1	4	11	17	21	25	28	27	22	13	5	0

    a) In which month is the average daily minimum temperature lowest? [1]

    b) In which month is the average daily maximum temperature highest? [1]

    c) In March, what is the difference between the average daily minimum temperature and the average daily maximum temperature? [1]

(PS) 2 The area of a square is 49cm². Work out the length of a side. [1]

Total Marks _____ / 4

(MR) 1 Jessa and Holly have been given the following question:

What is the value of $3 + 5 \times 4 + 7$?

Jessa thinks the answer is 30 and Holly thinks the answer is 39. Who is right? Explain your answer. [2]

(FS) 2 A netball club is planning a trip. The club has 354 members and the cost of the trip is £12 per member. 📱

    a) Work out the total cost of the trip. [3]

They need coaches for the trip and each coach seats 52 people.

    b) How many coaches do they need to book? [3]

    c) How many spare seats will there be? [2]

(MR) 3 Explain why $\sqrt{79}$ must be between 8 and 9. [2]

Total Marks _____ / 12

Sequences

1 **a)** Write down the next two numbers in the following sequence.

   3, 7, 11, 15, _____ , _____ ,... [2]

   b) Write down the term to term rule. [1]

2 Lynne plants a new flower in her garden.

   Of the buds, five have already flowered.

   Each week another three buds flower.

   a) How many buds will have flowered after three weeks? [1]

   b) How many weeks will it take for 32 buds to have flowered? [1]

Total Marks _____ / 5

1 **a)** Find the nth term of this arithmetic sequence.

   4, 7, 10, 13, 16,... [3]

   b) Find the 60th term in the sequence. [1]

(MR) **2** Match the cards on the left with a card on the right.

5, 9, 13, 17, 21...	Neither
2, 8, 18, 32, 50...	Quadratic
8, 17, 32, 53, 80...	Arithmetic

[2]

Total Marks _____ / 6

Perimeter and Area 1

You must be able to:

- Find the perimeter and area of a rectangle
- Find the area of a triangle
- Find the area and perimeter of compound shapes.

Perimeter and Area of Rectangles

- The **perimeter** is the distance around the outside of a 2D shape.
- The formula for the perimeter of a rectangle is:

    perimeter = 2(length + width) or $P = 2(l + w)$

  also   perimeter = 2(length) + 2(width)

length (l)

width (w)

- The formula for the **area** of a rectangle is:

  area = length × width or $A = l \times w$

Example

Find the perimeter and area of this rectangle.

8 cm

3 cm

Perimeter = 2(8 + 3)
= 2 × 11
= 22 cm
Area = 8 × 3
= 24 cm²

Area of a Triangle

- The formula for the area of a triangle is:

  area = $\frac{1}{2}$(base × **perpendicular** height)

Example

Find the area of this triangle.

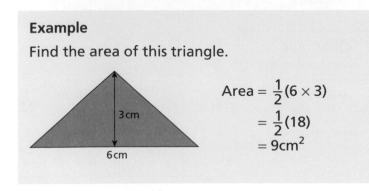

3 cm

6 cm

Area = $\frac{1}{2}$(6 × 3)

= $\frac{1}{2}$(18)

= 9 cm²

Key Point

When finding the area of a triangle, always use the perpendicular height.

Area and Perimeter of Compound Shapes

- A **compound** (or composite) shape is made up from other, simpler shapes.
- To find the area of a compound shape, divide it into basic shapes.

Example

This shape can be broken up into three rectangles.

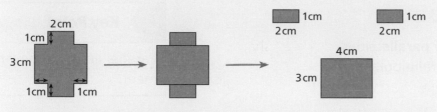

The areas of the individual rectangles are 2cm², 2cm² and 12cm².

The area of the compound shape is $2 + 2 + 12 = 16$cm².

> **Key Point**
>
> Areas are two-dimensional and are measured in square units, for example cm².

- To find the perimeter, start at one corner of the shape and travel around the outside, adding the lengths.

Example

Perimeter
$= 2 + 1 + 1 + 3 + 1 + 1 + 2 + 1 + 1 + 3 + 1 + 1$
$= 18$cm

Quick Test

1. Find the perimeter of a rectangle with width 5cm and length 7cm.
2. Find the area of a rectangle with width 9cm and length 3cm. Give appropriate units in your answer.
3. Find the area of a triangle with base 4cm and perpendicular height 3cm.
4. Find the perimeter and area of this shape.

> **Key Words**
>
> perimeter
> area
> perpendicular
> compound

Perimeter and Area 2

You must be able to:

- Find the area of a parallelogram
- Find the area of a trapezium
- Find the circumference and area of a circle.

Area of a Parallelogram

- A **parallelogram** has two pairs of **parallel** sides.
- The formula for the area of a parallelogram is:

area = base × perpendicular height

Example
Find the area of this parallelogram.

The base is 7cm and the perpendicular height is 3cm.

Area = 7 × 3
= 21cm^2

Area of a Trapezium

- A **trapezium** has one pair of parallel sides.
- The formula for the area of a trapezium is:

$A = \frac{1}{2}(a + b)h$

- The sides labelled a and b are the parallel sides and h is the perpendicular height.
- Perpendicular means at right angles.

Example
Find the area of this trapezium.

Area = $\frac{1}{2}(6 + 9) \times 4$
= 30cm^2

Circumference and Area of a Circle

- The **circumference** of a circle is the distance around the outside.
- The radius of a circle is the distance from the centre to the circumference.
- The diameter of a circle is twice the radius.
- The formula for the circumference of a circle is:
  $C = 2\pi r$ or $C = \pi d$
- The formula for the area of a circle is:
  $A = \pi r^2$

Key Point

The symbol π represents the number pi
(3.141 592 654...).

π is approximately 3.14 or $\frac{22}{7}$

Example

Find the circumference and area of this circle.
Give your answers to 1 decimal place.

7 cm

The circumference:

$C = 2 \times \pi \times 7$
$= 14 \times \pi$
$= 44.0 \text{cm}$

The area:

$A = \pi \times 7^2$
$= \pi \times 49$
$= 153.9 \text{cm}^2$

- Circles can be split into **sectors**.
- A sector is a region bounded by two radii and an arc (a curved line that is part of the circumference).
- A sector with a 90° angle at the centre would have an area of $\frac{1}{4}$ of the whole circle.

Sector

Quick Test

1. Find the area of the parallelogram.

8 cm

2 cm

2. Find the area of the trapezium.

5 cm

2 cm

8 cm

3. Find the circumference and area of a circle with radius 6cm.
4. Find the circumference and area of a circle with diameter 4cm.

Key Words

parallelogram
parallel
trapezium
pi (π)
circumference
sector

Statistics and Data 1

You must be able to:

- Find the mean, median, mode and range for a set of data
- Choose which average is the most appropriate to use in different situations
- Use a tally chart to collect data.

Mean, Median, Mode and Range

- The **mean** is the **sum** of all the values divided by the number of values.
- The **median** is the middle value when the data is in order.
- The **mode** is the most common value.
- The mean, median and mode are all averages.
- The **range** is the **difference** between the biggest and the smallest value.
- The range is a measure of spread.

> **Key Point**
>
> Data can have more than one mode. Bi-modal means the data set has two modes.

Example

Find the mean, median, mode and range for the following set of data:

$$5, 9, 7, 6, 2, 7, 3, 11, 4, 2, 6, 6, 4$$

The mean $= \dfrac{5+9+7+6+2+7+3+11+4+2+6+6+4}{13} = 5.5...$

Mean = 5.5 (1 d.p.)

The median = 2, 2, 3, 4, 4, 5,⑥ 6, 6, 7, 7, 9, 11

Median = 6

The mode is also 6 as this number is seen most often.

The range = 11 − 2 = 9

> Put the data in order, smallest to biggest.

Choosing which Average to Use

- Use the **mode** when you are interested in the most common answer, for example if you were a shoe manufacturer deciding how many of each size to make.
- Use the **mean** when your data does not contain **outliers**. A company which wanted to find average sales across a year would want to use all values.
- Use the **median** when your data does contain outliers, for example finding the average salary for a company when the manager earns many times more than the other employees.

> **Key Point**
>
> An outlier is a value that is much higher or lower than the others.

Constructing a Tally Chart

- A tally **chart** is a quick way of recording data.
- Your data is placed into groups, which makes it easier to analyse.
- A tally chart can be used to make a **frequency** chart by adding an extra column to record the total in each group.

Example

You are collecting and recording data about people's favourite flavour crisps. You ask 50 people and fill in the tally chart with their responses.

Flavour	Tally	Frequency
Plain	‖‖ ‖‖ ‖	12
Salt and vinegar	‖‖ ‖‖	9
Cheese and onion	‖‖ ‖‖ ‖‖ ‖	16
Prawn cocktail	‖‖ ‖	7
Other	‖‖ ‖	6
Total		**50**

From the table it can be seen that cheese and onion is the mode.

Quick Test

1. Emma surveyed her class to find out their favourite colour. She constructed the tally chart below.
   a) Complete the frequency column.
   b) How many pupils are there in Emma's class?

Colour	Tally	Frequency
Red	‖‖ ‖‖	
Blue	‖‖ ‖	
Green	‖‖ ‖	
Yellow	‖‖ ‖	
Other	‖‖ ‖	

2. Look at this set of data:
   3, 7, 4, 6, 3, 5, 9, 40, 6
   a) Write down the mode of this data.
   b) Calculate the mean, median and range.
   c) Would you choose the mean or the median to represent this data? Explain your answer.

Statistics and Data 2

You must be able to:

- Group data and construct grouped frequency tables
- Draw a stem-and-leaf diagram
- Construct and interpret a two-way table.

Grouping Data

- When you have a large amount of data it is sometimes appropriate to place it into groups.
- A group is also called a **class interval**.
- The disadvantage of using **grouped data** is that the original **raw data** is lost.

> **Example**
> The data below represents the number of people who visited the library each day over a 60-day period.
>
> | Number of people | Frequency |
> |:---:|:---:|
> | 0–10 | 10 |
> | 11–20 | 30 |
> | 21–30 | 14 |
> | Over 30 | 6 |
>
> On 30 out of the 60 days, the library had between 11 and 20 (inclusive) visitors.

- Data such as the number of people is discrete as it can only take particular values.
- Data such as height and weight is continuous as it can take any value on a particular scale.

Stem-and-Leaf Diagrams

- A stem-and-leaf diagram is a way of organising the data without losing the **raw data**.
- The data is split into two parts, for example tens and units.
- In the example below, the stem is the tens and the leaves the units.
- Each row should be ordered from smallest to biggest.
- A stem-and-leaf diagram must have a **key**.

```
        KEY: 1 | 2 = 12
  0 |  1 6 6 7 9 ── This is the number 9
  1 |  2 2 4 5 8 8 9
  2 |  3 2 5 5 7
  3 |  0 2 ── This is the number 32
```

Key Point

Calculations based on grouped data will be estimates.

Key Point

A stem-and-leaf diagram orders data from smallest to biggest.

Two-Way Tables

- A two-way table shows information that relates to two different categories.
- Two-way tables can be constructed from information collected in a survey.

Example

Daniel surveyed his class to find out if they owned any pets. In his class there are 16 boys and 18 girls. 10 of the boys owned a pet and 15 of the girls owned a pet.

	Pets	No Pets	Total
Boys	10		16
Girls	15		18
Total			

The information given is filled into the table and then the missing information can be worked out.

If there are 16 boys in Daniel's class and 10 of them have pets, then to work out how many boys do not have pets we calculate $16 - 10 = 6$

When all the missing information is entered you can **interpret** the data.

	Pets	No Pets	Total
Boys	10	6	16
Girls	15	3	18
Total	25	9	34

You can see that there are 34 pupils in Daniel's class and 25 of them owned a pet. You can also see that more girls owned pets than boys, and that more pupils owned pets than did not own pets.

> **Key Point**
>
> Two-way tables can also be used to estimate probabilities.

> **Quick Test**
>
> 1. 25 women and 30 men were asked if they preferred football or rugby. 16 of the women said they prefer football and 10 of the men said they prefer rugby.
>   a) Construct a two-way table to represent this information.
>   b) How many in total said they prefer football?
>   c) How many women preferred rugby?
>   d) How many people took part in the survey?

> **Key Words**
>
> class interval
> grouped data
> raw data
> key
> interpret

Review Questions

Number

FS **1** I purchase a new car for £3000.

I pay a deposit of £800 and pay the rest in four equal payments.

How much is each payment? [3]

2 The planets in our solar system orbit the Sun.

Their orbits are almost circular.

The table gives the time it takes, in days, for each
planet to complete one orbit of the Sun.

Earth	365
Jupiter	4332
Mars	687
Mercury	88
Neptune	60 200
Saturn	10 760
Uranus	30 700
Venus	224

Put the planets in order, starting with the shortest orbit time. [2]

3 A square has a perimeter of 20cm.

Work out the area of the square. [2]

4 Work out $4 \times 2^2 + 6$ [2]

Total Marks .... / 9

MR **1** $48 \times 52 = 2496$

Use this to help you work out the following calculations.

$24 \times 52 =$ .... $48 \times$ .... $= 1248$ $2496 \div 52 =$ .... [3]

PS **2** The lowest common multiple of two numbers is 60 and their sum is 27.

What are the numbers? [2]

FS **3** Gautam is having a barbecue and wants to invite some friends.

Sausages come in packs of 6. Rolls come in packs of 8.

She needs exactly the same number of sausages and rolls.

What is the minimum number of each pack she can buy? [3]

Total Marks .... / 8

Sequences

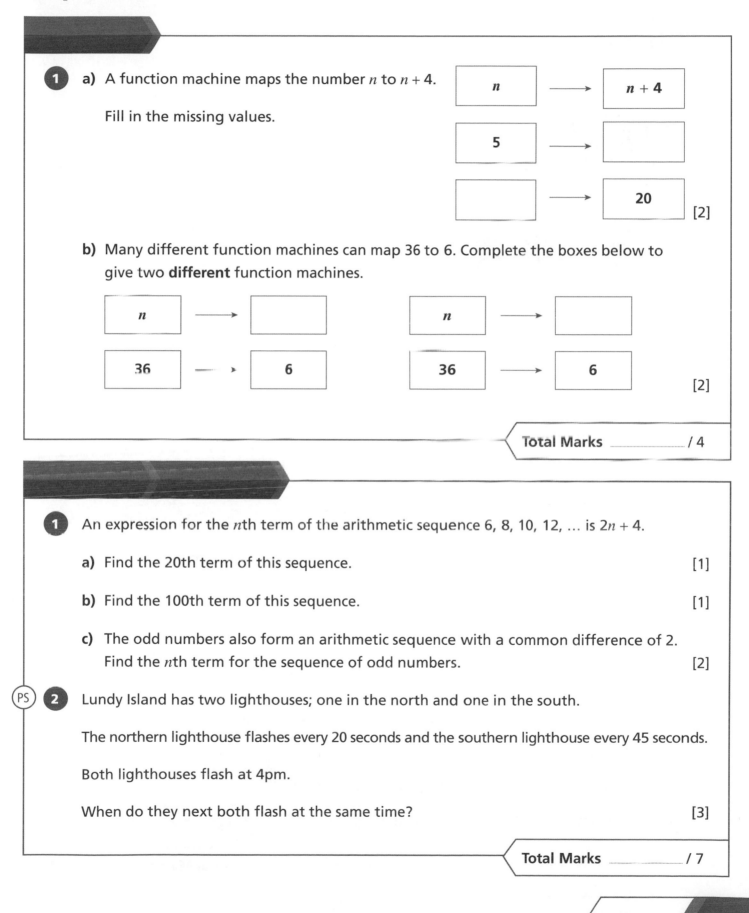

1 a) A function machine maps the number n to $n + 4$.

Fill in the missing values.

$n \longrightarrow n + 4$

$5 \longrightarrow $

$ \longrightarrow 20$ [2]

b) Many different function machines can map 36 to 6. Complete the boxes below to give two **different** function machines.

$n \longrightarrow $

$36 \longrightarrow 6$

$n \longrightarrow $

$36 \longrightarrow 6$ [2]

Total Marks _____ / 4

1 An expression for the nth term of the arithmetic sequence 6, 8, 10, 12, … is $2n + 4$.

a) Find the 20th term of this sequence. [1]

b) Find the 100th term of this sequence. [1]

c) The odd numbers also form an arithmetic sequence with a common difference of 2. Find the nth term for the sequence of odd numbers. [2]

(PS) 2 Lundy Island has two lighthouses; one in the north and one in the south.

The northern lighthouse flashes every 20 seconds and the southern lighthouse every 45 seconds.

Both lighthouses flash at 4pm.

When do they next both flash at the same time? [3]

Total Marks _____ / 7

Practice Questions

Perimeter and Area

1 Work out the perimeter and area of the rectangle below.

6cm

4cm

[2]

2

In the diagram above the area of the triangle is 8cm².

Use this to help you work out the area of the rectangle. [2]

Total Marks .... / 4

(PS) **1** The area of this rectangle is 48cm².

Find the value of X and Y.

Y cm

X cm

1.2cm

6cm

[2]

(FS) **2** Kelly is tiling her bathroom wall.

The wall is 4m by 3m.

Each tile is 25cm by 25cm.

a) Work out how many tiles Kelly needs to buy to tile the wall. [3]

The tiles come in packs of 10 and each pack costs £15.

b) Work out how much it will cost Kelly to tile the wall. [2]

c) How many tiles will she have left over? [1]

Total Marks .... / 8

Statistics and Data

1   Look at these five numbers: 6   11   9   12   7

     a)   Show that the mean of the five numbers is 9.   [1]

     b)   Explain why the median is also 9.   [1]

     c)   Write down a different set of five numbers which also have a mean of 9.   [1]

PS   2   The scores for a netball team are 15, 23, 8, 5, 19, x.

     The mean score is 15. What is x?   [2]

3   This stem-and-leaf diagram shows the marks of some students in a test.

4	1 5 6 7 8 9
5	2 3 4 6 8 9 9 9
6	3 4 5 5 8 9
7	0 2 6

KEY: 5 | 2 = 52 marks

     A student was absent on the day the test was taken, took it at a later date and scored 78.

     a)   Add this score to the stem-and-leaf diagram.   [1]

     b)   Write down the mode.   [1]

     c)   Work out the range of the scores.   [1]

Total Marks .... / 8

MR   1   Phil and Dave are both good darts players.

     Their scores for a match are shown below.

Phil	64	70	80	100	57	100	41	56	30
Dave	36	180	21	180	10	5	23	25	140

     a)   Calculate the mean score for each player.   [2]

     b)   Find the range of scores for each player.   [2]

     c)   Only one of the two players can be picked to play in the next match.

        Would you pick Phil or Dave? Explain your answer.   [2]

Total Marks .... / 6

Algebra 1

You must be able to:

- Know the difference between an equation and expression
- Collect like terms in an expression
- Write products as algebraic expressions
- Substitute numerical values into formulae and expressions.

Collecting Like Terms

- The difference between an **equation** and an **expression** is that an equation has an equals sign.
- To **simplify** an expression like terms are collected.

Example

Simplify $2x + 6y - x + 4y$

Collect the like terms:

$2x - x + 6y + 4y$

The x terms can be simplified: $2x - x = x$

The y terms can be simplified: $6y + 4y = 10y$

So $2x + 6y - x + 4y$ can be simplified to $x + 10y$

Example

Simplify $2x^2 + 6y - x^2 + 4y - 6$

Collect the like terms:

$2x^2 - x^2 + 6y + 4y - 6$

The x terms can be simplified: $2x^2 - x^2 = x^2$

The y terms can be simplified: $6y + 4y = 10y$

There is only one constant term.

So $2x^2 + 6y - x^2 + 4y - 6$ can be simplified to $x^2 + 10y - 6$

Example

Simplify $\frac{2}{3}x + y - \frac{1}{3}x + \frac{3}{4}y$

Collect the like terms:

$\frac{2}{3}x - \frac{1}{3}x + y + \frac{3}{4}y$

The x terms can be simplified: $\frac{2}{3}x - \frac{1}{3}x = \frac{1}{3}x$

The y terms can be simplified: $y + \frac{3}{4}y = \frac{7}{4}y$

So $\frac{2}{3}x + y - \frac{1}{3}x + \frac{3}{4}y$ can be simplified to $\frac{1}{3}x + \frac{7}{4}y$

> **Key Point**
>
> Remember that terms have a + or − sign between them and each sign belongs to the term on its right.

Expressions with Products

- **Product** means multiply.
- Expressions with products are written without the $\times$ sign.

> **Example**
>
> $2 \times a = 2a$
>
> $a \times b = ab$
>
> $a \times a \times b = a^2b$
>
> $a \div b = \dfrac{a}{b}$
>
> $a \times a \times a = a^3$

Substitution

- A **formula** is a rule which links a variable to one or more other variables.
- The variables are written in shorthand by representing them with a letter.
- Some commonly used scientific formulae are:

$$\text{speed} = \frac{\text{distance}}{\text{time}} \qquad \text{in shorthand } s = \frac{d}{t}$$

$$\text{density} = \frac{\text{mass}}{\text{volume}} \qquad \text{in shorthand } d = \frac{m}{v}$$

- Substitution involves replacing the letters in a given formula or expression with numbers.

> **Example**
>
> Find the value of the expression $2a + b$ when $a = 3$ and $b = 5$.
>
> Replace the letters with the given numbers.
>
> $2a + b = 2 \times 3 + 5 = 11$

> **Example**
>
> Emma's mum lives 30 miles from her and on a particular morning last week Emma's journey there took $\frac{3}{4}$ of an hour.
>
> Calculate her average speed in miles per hour.
>
> $s = \dfrac{d}{t}$
>
> $s = \dfrac{30}{0.75} = 40\text{mph}$

 $\frac{3}{4} = 0.75$

> **Key Point**
>
> There are many scientific problems which involve substituting into formulae. Always follow BIDMAS.
>
> To find the area and volume of shapes, we substitute into a formula.

> **Quick Test**
>
> 1. Simplify $4x + 7y + 3x - 2y + 6$
> 2. Simplify $c \times c \times d \times d$
> 3. Find the value of $4x + 2y$ when $x = 2$ and $y = 3$

> **Key Words**
>
> equation
> expression
> simplify
> product
> formula

Algebra 2

You must be able to:

- Multiply a single term over a bracket
- Factorise linear expressions.

Expanding Brackets

- **Expanding** the brackets involves removing the brackets by **multiplying** every term inside the bracket by the number or term on the outside.
- This table is a reminder of the rules when multiplying and dividing negative numbers.

	+	−
+	+	−
−	−	+

Example

Expand $3(x + 5)$

×	3
x	$3x$
+5	+15

$3(x + 5) = 3x + 15$

Key Point

Make sure you multiply every term inside the bracket.

Example

Expand and simplify $4(x + y) - 2(2x - 3y)$

×	4
x	$4x$
y	$4y$

$= 4x + 4y$

×	2
$2x$	$4x$
$-3y$	$-6y$

$= 4x - 6y$

Then collect like terms.

$(4x + 4y) - (4x - 6y) = 10y$

Factorising

- **Factorising** is the opposite of expanding.
- When factorising we put the brackets back in.

Example

Factorise $6x + 9$

3 is a common factor of 6 and 9 so we take the 3 to the outside of the bracket.

To find what is inside the bracket we need to fill in the blanks in the table.

×	3
	$6x$
	$+9$

×	3
$2x$	$6x$
$+3$	$+9$

So $6x + 9 = 3(2x + 3)$

Key Point

Always expand your answer to check you are right.

Example

Factorise $x^2 + 2x$

x is the common factor:

×	x
	x^2
	$+2x$

×	x
x	x^2
$+2$	$+2x$

So $x^2 + 2x = x(x + 2)$

Example

Factorise $6x^3 + 2x^2$

2 and x^2 are the common factors:

×	$2x^2$
	$6x^3$
	$2x^2$

×	$2x^2$
$3x$	$6x^3$
$+1$	$2x^2$

So $6x^3 + 2x^2 = 2x^2(3x + 1)$

Quick Test

1. Expand $4(2x - 1)$
2. Expand and simplify $2(2x - y) - 2(x + 6y)$
3. Factorise $5x - 25$
4. Factorise completely $2x^2 - 4x$

Key Words

expand
factorise

Review Questions

Perimeter and Area

1. Find the area of a circle with radius 7cm. [2]

2. Find the perimeter and area of the rectangle below.

   5cm

   9cm

   [2]

3. Find the area of the trapezium below.

   6cm

   4cm

   8cm

   [2]

Total Marks _____ / 6

1. Frances wants to paint the front of her house.
   The diagram represents the front of her house.

   a) Find the area which Frances needs to paint. [3]

   Each tin of paint covers 7m².

   b) Work out the number of tins which Frances needs to buy. [1]

   Each tin costs £12.

   c) Work out how much it will cost Frances to paint the front of her house. [2]

   6m
   5m
   4m

(PS) 2. Edmund buys a new bicycle and uses it to cycle 8000m to work every day and then catches the train home.

   His wheel is a circle with a diameter of 50cm.

   Work out how many times his wheel makes a full rotation during his journey. [2]

Total Marks _____ / 8

Statistics and Data

1 There are 34 pupils in class 8D and 19 of those are boys.

26 of the class are right-handed and 14 of those are girls.

Use this information to complete the table below. [3]

	Boys	Girls
Right-handed		
Left-handed		

2 James does a survey of the favourite vegetables of the pupils in his class:

carrots	potatoes	peas	sweetcorn	peas	peas
potatoes	peas	carrots	peas	potatoes	sweetcorn
potatoes	potatoes	potatoes	carrots	carrots	potatoes
peas	potatoes	sweetcorn	peas	potatoes	peas

a) Construct a frequency table to represent this data. [2]

b) What was the mode? [1]

Total Marks _____ / 6

MR **1** This data below shows the number of people attending six matches for Sandex United.

1240   1354   1306   14808   1378   1430

You want to calculate the average attendance.

a) Would you find the mean, median or mode? Give a reason for your answer. [2]

b) Which of the attendance numbers is an outlier? [1]

Total Marks _____ / 3

Practice Questions

Decimals

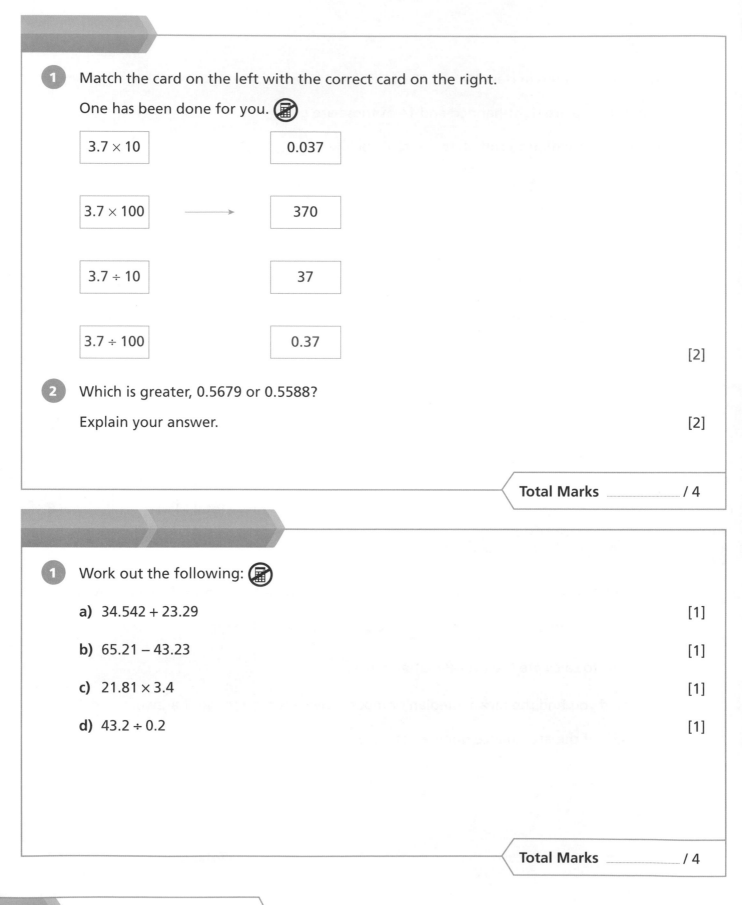

1. Match the card on the left with the correct card on the right.
   One has been done for you.

3.7 × 10		0.037
3.7 × 100	→	370
3.7 ÷ 10		37
3.7 ÷ 100		0.37

   [2]

2. Which is greater, 0.5679 or 0.5588?

   Explain your answer.   [2]

   Total Marks _____ / 4

1. Work out the following:

   a) 34.542 + 23.29   [1]

   b) 65.21 − 43.23   [1]

   c) 21.81 × 3.4   [1]

   d) 43.2 ÷ 0.2   [1]

   Total Marks _____ / 4

Algebra

1 Complete the statements.

When $x = 5$ $5x = $ _____

When $x = $ _____ $5x = 45$

When $x = 7$ _____ $= 28$ [2]

2 In this algebra grid, the total on each brick is made from the sum of the two bricks below. The first brick has been calculated for you. Complete the grid.

$3y + 2y = 5y$ →

5y
3y	2y	5y

[2]

Total Marks _____ / 4

MR 1 Claudia thinks the perimeter of this rectangle is $2x + 2y$.
Lawrence thinks the perimeter of the rectangle is $2(x + y)$.

Who is right – Claudia, Lawrence or both of them?
Explain your answer. [2]

FS 2 The cost (in £) of hiring a car for the day and driving it y miles is shown by this formula:

$C = 75 + 0.4y$

Work out how much it would cost to hire the car for a day and travel 120 miles. [3]

3 Expand and simplify $2(4x + 1) - 5(x - 1)$. [2]

4 Factorise completely $3abc + 6a$. [2]

PS 5 If $ab = 36$ and $a = 4$, find the value of a^2b. [2]

6 Complete the algebra grid below. The total in each brick is made from the sum of the two bricks below.

5a − b	
	2a
a		

[2]

Total Marks _____ / 13

3D Shapes: Volume and Surface Area 1

You must be able to:

- Name and draw 3D shapes
- Draw the net of a cuboid and other 3D shapes
- Calculate the surface area and volume of a cuboid.

Naming and Drawing 3D Shapes

- A 3D shape can be described using the number of **faces**, **vertices** and **edges** it has.

Shape	Name	Edges	Vertices	Faces
	Cuboid	12	8	6
	Triangular prism	9	6	5
	Square-based pyramid	8	5	5
	Cylinder	2	0	3
	Pentagonal prism	15	10	7

> **Key Point**
>
> A face is a sur'face', for example a flat side of a cube.
>
> A vertex is where edges meet, for example the corner of a cube.
>
> An edge is where two faces join.

Using Nets to Construct 3D Shapes

- To create the **net** of a cuboid, imagine it is a box you are unfolding to lay out flat.

- Nets for a triangular prism and a square-based pyramid look like this:

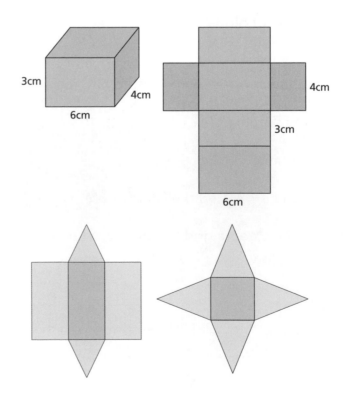

3cm
4cm
6cm
4cm
3cm
6cm

Surface Area of a Cuboid

- The **surface area** of a cuboid is the sum of the areas of all six faces and is measured in square units (cm², m², etc.).

Example

To calculate the surface area of the cuboid on the previous page, you can use the net to help you.

Work out the area of each rectangle by multiplying the base by its height:

Green rectangle: $6cm \times 4cm = 24cm^2$
There are two of them, so $24cm^2 \times 2 = 48cm^2$

Blue rectangle: $3cm \times 6cm = 18cm^2$
There are two of them, so $18cm^2 \times 2 = 36cm^2$

Pink rectangle: $4cm \times 3cm = 12cm^2$
There are two of them, so $12cm^2 \times 2 = 24cm^2$

Sum of all six areas is $48 + 36 + 24 = 108cm^2$

Volume of a Cuboid

- **Volume** is the space contained inside a 3D shape.

Example

To calculate the volume of the cuboid on the previous page, you need to first work out the area of the front rectangle:

$3cm \times 6cm = 18cm^2$

Next multiply this area by the depth of the cuboid, 4cm.

$18cm^2 \times 4cm = 72cm^3$

⟵ The units are cm³ this time.

Quick Test

Work out the volume and the surface area of these cuboids.

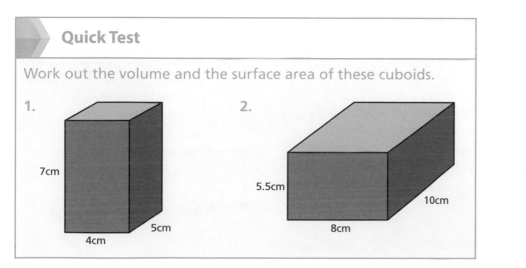

1. 7cm, 4cm, 5cm

2. 5.5cm, 8cm, 10cm

Key Words

face
vertex
edge
prism
net
surface area
volume

3D Shapes:
Volume and Surface Area 2

You must be able to:

- Calculate the volume of a cylinder
- Calculate the surface area of a cylinder
- Calculate the volume of composite shapes.

Volume of a Cylinder

- You work out the volume of a **cylinder** the same way as the volume of a cuboid. First work out the area of the **circle** and then multiply it by the height of the cylinder.

Example

Calculate the volume of this cylinder.

Volume = $(\pi \times 5 \times 5) \times 3 = 235.62\text{cm}^3$
(2 d.p.)

This cylinder has a **diameter** of 12cm.

This means the **radius** is 6cm.

Volume = $(\pi \times 6 \times 6) \times 5 = 565.49\text{cm}^3$
(2 d.p.)

> **Key Point**
>
> Diameter is the full width of a circle that goes through the centre.
>
> Radius is half of the diameter.

Area of a circle = $\pi \times$ radius²

Volume units are shown by a ³, for example cm³.

Surface Area of a Cylinder

- To calculate the surface area, first draw the net of the cylinder (imagine cutting a can open).

Example

Calculate the surface area of this cylinder.

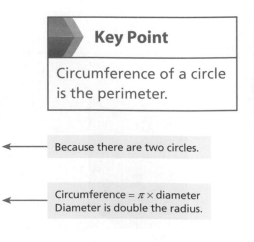

The net is made up from two circles and a rectangle.
You need to find the area and circumference of the circles.
Calculate the area of the circular ends: $(\pi \times 3 \times 3) \times 2 = 56.548...$
The area of the rectangle is the circumference of the circle multiplied by the height of the cylinder, 10cm.

$(\pi \times 6) \times 10 = 188.495...$

Add the areas together: $56.548... + 188.495... = 245.04\text{cm}^2$ (2 d.p.)

> **Key Point**
>
> Circumference of a circle is the perimeter.

Because there are two circles.

Circumference = $\pi \times$ diameter
Diameter is double the radius.

Calculating the Volume of Composite Shapes

- **Composite** means the shape has been 'built' from more than one shape.

Example

This shape is built from two cuboids.

Calculate the volume of the two separate cuboids and add the volumes together.

$(6cm - 2cm) \times 5 \times 3 = 60cm^3$

$(3cm + 4cm) \times 2 \times 5 = 70cm^3$

$60 + 70 = 130cm^3$

Quick Test

1. Work out the volume and the surface area of these shapes.

   a)

   b)

2. Work out the volume of this composite shape.

Key Words

cylinder
circle
diameter
radius
composite

Interpreting Data 1

You must be able to:

- Create a simple pie chart from a set of data
- Create and interpret pictograms
- Draw a frequency diagram
- Make comparisons and contrasts between data.

Pie Charts

- **Pie charts** are often shown with **percentages** or **angles** indicating sector size – this and the visual representation helps to **interpret** the **data**.

Example

36 students were asked the following question:

Which is your favourite flavour of crisps?

To work out the angle for each sector: 360 ÷ 36 = 10°

$$\frac{\text{Degrees in full turn}}{\text{Total}} = \text{Degrees per person}$$

Flavour of crisps	No. of students	Degrees
Salt and Vinegar ■	8	8 × 10 = 80
Cheese and Onion ■	10	10 × 10 = 100
Ready Salted ■	12	12 × 10 = 120
Prawn Cocktail ■	4	4 × 10 = 40
Other ■	2	2 × 10 = 20

Key Point

Always align your protractor's zero line with your starting line.

Count up from zero to measure your angle.

Don't forget to label your chart.

Pictograms

- Data is represented by a picture or symbol in a **pictogram**.

Example

The pictogram shows how many pizzas were delivered by Ben in one week. Key 🍕 = 8 pizzas

Day	Mon	Tue	Wed	Thu	Fri	Sat	Sun
Pizza deliveries	🍕🍕🍕	🍕	🍕🍕	🍕	🍕🍕🍕🍕	🍕🍕	

How many pizzas did Ben deliver on Friday? 8 × 4 = 32

On Sunday Ben delivered 20 pizzas. Complete the pictogram.

20 ÷ 8 = 2.5 🍕🍕🍕

Frequency Diagrams

- Frequency diagrams are used to show grouped data.

Example

The heights of 100 students are shown in this table.

Height (*h*) in cm	Frequency
$70 < h \leqslant 80$	3
$80 < h \leqslant 90$	4
$90 < h \leqslant 100$	12
$100 < h \leqslant 110$	24
$110 < h \leqslant 120$	30
$120 < h \leqslant 130$	22
$130 < h \leqslant 140$	3
$140 < h \leqslant 150$	2

Using the span of each height category, plot each group as a block using the frequency axis.

Data Comparison

- You can make comparisons using frequency diagrams.

Example

These graphs show how much time four students spend on their mobile phones in a week. Compare the two graphs.

Apart from Thursday, Helen uses her phone more than Andy. In Ian's and Rhian's data, Rhian uses her phone more overall. Comparing both sets of data we can say the girls use their phones more than the boys.

Key Point

Think about what is similar about the data and what is different. Look for any patterns.

Quick Test

1. If 18 people were asked a question and you were to create a pie chart to represent your data, what angle would one person be worth?
2. Using the graph above right, on which day did both Helen and Andy use their mobiles the most?
3. The following week Andy had a mean use of 40 minutes per day. Does this mean he uses his phone more than Helen now?

Key Words

pie chart
percentage
angle
interpret
data
pictogram
frequency
axis

Interpreting Data 2

You must be able to:

- Interpret different graphs and diagrams
- Draw a scatter graph and understand correlation
- Understand the use of statistical investigations.

Interpreting Graphs and Diagrams

- You can interpret the information in graphs and diagrams.

Example

What does this graph show?

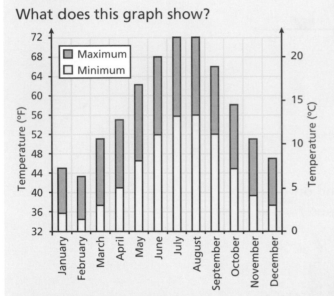

Using the labels and key, you can see that it gives the temperature and months of a year, plus a maximum and minimum temperature.

Which month has the lowest temperature?

Using the minimum temperature (yellow bars), pick the smallest bar, in this case, February.

Example

Look at this pie chart. What is the most likely way the team scores a goal?

Scoring a goal in free play represents the largest sector of the pie chart.

Goals scored in a season

- Own goals 2%
- Penalties 10%
- Corners 29%
- Free play 39%
- Free kicks 20%

Drawing a Scatter Graph

- A **scatter graph** plots two bits of information on one graph.
- A scatter graph can show a potential relationship between data.
- You can draw a **line of best fit** through the points on a scatter graph.

Example

Here is a plot of the sales of ice cream against the amount of sun per day for 12 days. The scatter graph shows that when the weather is sunnier (and hotter), more ice creams are bought.

Key Point

Plot each data pair as a coordinate. A line of best fit doesn't have to start at zero.

- You can describe the **correlation** and use the line of best fit to estimate data values.
- The graph above shows a positive correlation between ice cream sales and sunlight in a day.

Example

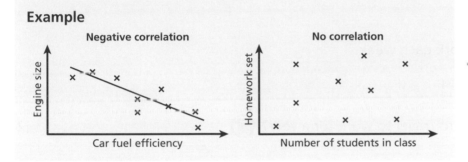

Statistical Investigations

- Statistical investigations use **surveys** and experiments to test statements and theories to see whether they might be true or false. These statements are called **hypotheses**.
- Survey questions should be specific, timely and have no overlap on answer choices.

Key Point

Surveys should be made on a large random sample, never just a limited few.

Quick Test

1. Name two things you might plot against one another to show a positive correlation.
2. Look at the pie chart on page 48. What is the least likely way that the team scored a goal?
3. Give two ways of improving this survey question.
   How many sweets do you eat?

   None ☐   1–2 ☐   2–3 ☐   3–4 ☐
4. Rewrite the question in number 3 to improve it.

Key Words

scatter graph
line of best fit
correlation
survey
hypothesis

Review Questions

Decimals

1. Work out 2.53×0.3 🖩 [2]

2. Work out $0.85 \div 0.05$ 🖩 [2]

3. Circle the two numbers which add to make 9. 🖩

   8.1   0.7   5.2   3.8   0.8 [1]

4. By rounding both numbers to 1 significant figure, find an estimate for the following: 🖩

   $\dfrac{6782}{53}$ [2]

5. I pay £17.80 to travel to work each week.

   I work for 48 weeks a year.

   How much does it cost me to travel to work for a year? 🖩 [2]

   Total Marks .... / 9

---

1. Put the following numbers in order from smallest to biggest. 🖩

   7.765, 7.675, 6.765, 7.756, 6.776 [2]

2. Thomas buys three books that cost £2.98, £3.47 and £9.54 🖩

   a) How much did the books cost in total? [2]

   b) How much change did he get from a £20 note? [1]

3. Use your calculator to work out $\sqrt{48}$ to 2 decimal places. [1]

4. A number has been rounded to 35 to the nearest whole number. Express the rounding error as an inequality. [1]

   Total Marks .... / 7

Algebra

(PS) **1** Simplify these expressions:

$5k + 5 + 6k$

$k + 2 + 3k - 1$ [2]

2 Fill in the missing term in the statement below.

$3k + 4$ _____ $= k + 4$ [1]

3 Complete the missing information in the table below. The first row has been done for you.

$c \times d$	cd
$c \times d \times d$	
$c \times c \times d$	
$c \times c \times d \times d$	

[2]

Total Marks _____ / 5

1 This rectangle has dimensions $a \times b$.

a) Write a simplified expression for the area of the rectangle. [1]

b) Write a simplified expression for the perimeter of the rectangle. [1]

c) Another rectangle has an area of $15a^2$ and a perimeter of $16a$. What are the dimensions of this rectangle? [1]

2 Factorise the following expression completely.

$8ut^2 - 4ut + 20t$ [2]

3 Lucy walked 1200 metres in 20 minutes.

$s = \dfrac{d}{t}$

Use the formula to find her average speed in metres per hour. [2]

Total Marks _____ / 7

Practice Questions

3D Shapes: Volume and Surface Area

1. Complete the table below.

Shape	Name	No. of faces	No. of edges	No. of vertices

[3]

2. Name the 3D solid for each net.

[3]

3. Calculate the volume and surface area of these two cuboids.

a) 10 cm, 3 cm, 5 cm

b) 4.5 cm, 3 cm, 6 cm

[4]

Total Marks _____ / 10

(MR) 1. Find the height of the cylinder with a radius of 5cm and a volume of 942cm³. Give your answer to 1 decimal place. [2]

(MR) 2. Find the height of the cylinder with a radius of 7cm and a volume of 1385cm³. Give your answer to 2 decimal places. [2]

Total Marks _____ / 4

Interpreting Data

(MR) **1** 45 people were asked what their favourite cheese was.

Complete the table and plot the information on a pie chart.

Category	Frequency	Angle
Brie	21	
Cheddar	5	
Stilton	14	
Other	5	
Total		

[6]

(MR) **2** The graph shows Helen and Andy's phone use during a week. On which day was their total amount of phone usage the lowest?

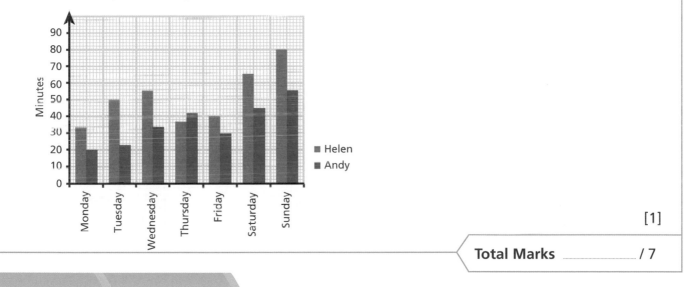

[1]

Total Marks .... / 7

1 What two things could you plot against one another to show a negative correlation? [2]

2 Write a question with response box options to determine whether people shop more over the Christmas period than at other times of the year. [4]

(MR) **3** a) Name four different types of statistical graphs or charts. [2]

b) Which one would you use to plot the information collected from asking 40 students, 'How long do you spend doing homework in a week?' Give a reason. [2]

Total Marks .... / 10

Fractions 1

You must be able to:

- Find equivalent fractions
- Add and subtract fractions.

Equivalent Fractions

- **Equivalent** fractions are fractions that are equal despite the **denominators** being different.

Example

$\frac{1}{2}$ $\frac{2}{4} = \frac{1}{2}$ $\frac{4}{8} = \frac{1}{2}$

- You can create an equivalent fraction by keeping the ratio between the **numerator** and denominator the same.
- You do this by multiplying or dividing both the numerator and denominator by the same number.
- Creating equivalent fractions is very useful when you want to compare or evaluate different fractions.

Example

Which is the larger fraction, $\frac{2}{5}$ or $\frac{3}{7}$?

$\frac{2}{5}$ $\frac{3}{7}$

To compare these fractions, you need to find a common denominator – a number that appears in both the 5 and 7 times tables.

$5 \times 7 = 35$

So $\frac{2}{5}$ becomes $\frac{14}{35}$ and $\frac{3}{7}$ becomes $\frac{15}{35}$

Now the denominators are equal, you can evaluate the two fractions more easily and you can see that $\frac{15}{35} = \frac{3}{7}$ is larger.

> **Key Point**
>
> A common denominator is a number that shares a relationship with both fractions' denominators.
>
> For example, for 5 and 3 this would be 15, 30, 45, 60, …

Adding and Subtracting Fractions

- Adding fractions with the same denominator is straightforward. The 'tops' (numerators) are collected together.

Example

$$\frac{1}{3} \quad + \quad \frac{1}{3} \quad = \quad \frac{2}{3}$$

Notice the size of the 'piece', the denominator, remains the same in both the question and the answer.

- When subtracting fractions with the same denominator, simply subtract one numerator from the other.

Example

$$\frac{4}{5} \quad - \quad \frac{1}{5} \quad = \quad \frac{3}{5}$$

> **Key Point**
>
> The size of the 'piece' (the denominator) has to be the same to perform either addition or subtraction.

- When you have fractions with different denominators, first find equivalent fractions with a common denominator.

Example

$$\frac{1}{4} \quad + \quad \frac{2}{3} \quad = \quad \frac{3}{12} + \frac{8}{12} = \frac{11}{12}$$

Here the common denominator is 12, as it is the smallest number that appears in both the 3 and 4 times tables.

This means that, for the first fraction, you have to multiply both the numerator and denominator by 3 and for the second fraction multiply them by 4.

Now the fractions are of the same size 'pieces', you can add the numerators as before.

> **Key Point**
>
> It is essential to find equivalent fractions so both fractions have the same denominator.

> **Quick Test**
>
> 1. Find three equivalent fractions for $\frac{2}{3}$
>
> 2. Work out $\frac{2}{7} + \frac{6}{11}$
>
> 3. Work out $\frac{7}{9} - \frac{3}{8}$
>
> 4. Work out $\frac{7}{13} - \frac{1}{4}$
>
> 5. Work out $\frac{14}{25} + \frac{3}{5} - \frac{7}{20}$

> **Key Words**
>
> equivalent
> denominator
> numerator

Fractions 2

You must be able to:

- Multiply and divide fractions
- Understand mixed numbers and improper fractions
- Calculate sums involving mixed numbers.

Multiplying and Dividing Fractions

- Multiplying fractions by whole numbers is not very different from multiplying whole numbers.
- The numerator is multiplied by the whole number.

> **Example**
>
> $\frac{2}{7} \times 3 = \frac{2}{7} + \frac{2}{7} + \frac{2}{7} = \frac{6}{7}$

> **Key Point**
>
> No common denominator is needed for multiplying or dividing fractions.

- When multiplying, you multiply the 'tops' (numerators) and then multiply the 'bottoms' (denominators).

> **Example**
>
> $\frac{3}{5} \times \frac{2}{7} = \frac{3 \times 2}{5 \times 7} = \frac{6}{35}$
>
> When you multiply these two fractions, it is like saying there are $\frac{3}{5}$ lots of $\frac{2}{7}$.

- When dividing, you use **inverse** operations. You change the operation to a multiplication and invert the second fraction.

> **Example**
>
> $\frac{2}{5} \div \frac{1}{2} = \frac{2 \times 2}{5 \times 1} = \frac{4}{5}$
>
> When you divide fractions, you are saying how many of one of the fractions is in the other. $\frac{2}{5}$ divided into halves gives twice as many pieces.

Mixed Numbers and Improper Fractions

- A **mixed number** is where there is both a whole number part and a fraction, for example $1\frac{1}{3}$
- An **improper fraction** is where the numerator is bigger than the denominator, for example $\frac{4}{3}$

Example

Change the improper fraction $\frac{14}{3}$ to a mixed number.

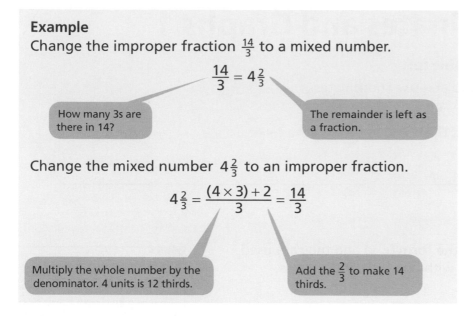

$$\frac{14}{3} = 4\frac{2}{3}$$

How many 3s are there in 14?

The remainder is left as a fraction.

Change the mixed number $4\frac{2}{3}$ to an improper fraction.

$$4\frac{2}{3} = \frac{(4 \times 3) + 2}{3} = \frac{14}{3}$$

Multiply the whole number by the denominator. 4 units is 12 thirds.

Add the $\frac{2}{3}$ to make 14 thirds.

Key Point

Improper fractions can be changed into a mixed number and vice versa.

Adding and Subtracting Mixed Numbers

* To work with mixed numbers, you first have to change them to improper fractions.
* There are four steps:
  convert to improper fractions → convert to equivalent fractions → add → convert back to a mixed number

Example

$$2\frac{4}{9} + 3\frac{1}{4} = \frac{22}{9} + \frac{13}{4} = \frac{22 \times 4}{36} + \frac{13 \times 9}{36} = \frac{205}{36} = 5\frac{25}{36}$$

* The same sequence of steps is needed for subtraction:
  convert to improper fractions → convert to equivalent fractions → subtract → convert back to a mixed number

Example

$$4\frac{4}{7} - 1\frac{1}{4} = \frac{32}{7} - \frac{5}{4} = \frac{32 \times 4}{28} - \frac{5 \times 7}{28} = \frac{93}{28} = 3\frac{9}{28}$$

Key Point

Always write the 'remainder' as a fraction.

Quick Test

Work out:

1. $\frac{4}{5} \times \frac{5}{12}$

2. $\frac{7}{12} \div \frac{3}{7}$

3. $\frac{12}{15} \div \frac{5}{35}$

4. $5\frac{5}{6} + 3\frac{5}{12}$

5. $4\frac{3}{10} - 2\frac{1}{9}$

Key Words

inverse
mixed number
improper fraction

Coordinates and Graphs 1

You must be able to:

- Understand and use coordinates
- Plot linear graphs
- Understand the components of $y = ax + b$.

Coordinates

- Coordinates are usually given in the form (x, y) and they are used to find certain points on a graph with an x-**axis** and a y-axis.

Example
Plot the coordinates (4, 7) and (–3, 4).

(4, 7) is 4 across and 7 up from 0.

> **Key Point**
>
> The x-axis is always the one that goes across the page. The y-axis always goes up the page.
>
> When reading or plotting a coordinate, we use the first number as the position on the x-axis and the second number as the position on the y-axis.

- You use the same idea to read a coordinate from a graph or chart.

Example
What are the coordinates of each corner of this square?

$A = (3, -2)$, $B = (6, -2)$, $C = (6, -5)$, $D = (3, -5)$

Linear Graphs

- **Linear** graphs form a straight line.
- When plotting a graph, we need to have a rule for what we are plotting, usually an equation.

Example

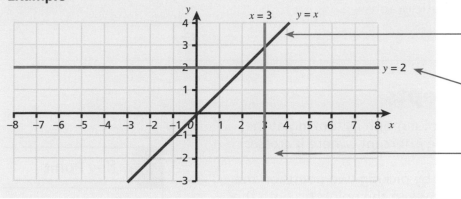

When plotting $y = x$, whatever x-coordinate you choose, the y-coordinate will be the same.

When plotting $y = 2$, you can choose any value for the x-coordinate but the y-coordinate must always equal 2.

When plotting $x = 3$, you can choose any value for the y-coordinate but the x-coordinate must always equal 3.

Graphs of $y = ax + b$

- a is used to represent the **gradient** of the graph.
- b is used to represent the **intercept**.
- To create the graph you substitute real numbers for x and y.

Example
Plot the graph $y = 2x + 1$

If $x = 1$, you can work out y: $y = 2 \times 1 + 1 = 3$
Work out other values of y by changing the value of x.

x	−1	0	1	2	3
y	−1	1	3	5	7

> **Key Point**
>
> A positive value of a will give a positive gradient. The graph will appear 'uphill' from left to right.
>
> A negative value of a will give a negative gradient. The graph will appear 'downhill' from left to right.

> **Quick Test**
>
> 1. Complete the table of values for $y = 3x - 5$.
>
> | x | −2 | −1 | 0 | 1 | 2 | 3 |
> |---|----|----|---|---|---|---|
> | y | | | | | | |

> **Key Words**
>
> coordinates
> axis
> linear
> gradient
> intercept

Coordinates and Graphs 2

You must be able to:

* Understand gradients and intercepts
* Solve equations from linear graphs
* Plot quadratic graphs.

Gradients and Intercepts

* You can work out the equation of a graph by looking at the gradient (how steep it is) and the intercept (where it crosses the y-axis).
* The gradient can be worked out by picking two points on the graph, finding the difference between the points on both the y- and x-axes and dividing them.

Key Point

gradient =

$$\frac{\text{difference in } y}{\text{difference in } x}$$

Example

When x increases by 1, y increases by 3.

$3 \div 1 = 3$ so the gradient is 3.

The line crosses the y-axis at –1 so the intercept is –1.

The equation is $y = 3x - 1$

Solving Linear Equations from Graphs

* Graphs can be used to solve linear equations visually.

Example
You can find the solution to the equation $2x + 3 = 7$ by using the graph $y = 2x + 3$

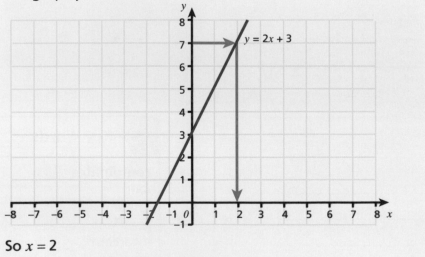

$y = 2x + 3$

So $x = 2$

First plot the graph. Then find where $y = 7$ on the y-axis. Trace your finger across until it meets the graph. Finally follow it down to read the x-axis value.

Key Point

You can check your solution from a graph by substituting back into the equation.

For example, if $x = 2$
$y = 2x + 3 = 2(2) + 3 = 7$

- **Simultaneous equations** are two equations that are linked.
- The solution to both equations can be seen at the point where their graphs cross.

Drawing Quadratic Graphs

- **Quadratic equations** make graphs that are not linear but curved.

Example

The equation $y = x^2$ has a power in it, so this alters the graph to one that has a curve.

x	–3	–2	–1	0	1	2	3
$y = x^2$	9	4	1	0	1	4	9

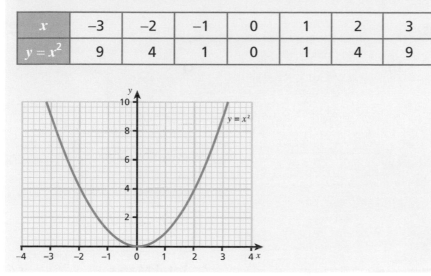

> **Key Point**
>
> Remember that when multiplying a negative by another negative, it becomes a positive number.

- Quadratics can take more complicated forms, but you still just **substitute** x for a real number to get the coordinates.

Example

$y = x^2 + 2x + 1$

If $x = -3$

$y = (-3)^2 + 2(-3) + 1$

$\quad = 9 + -6 + 1 = 4$

Work out other values of y by changing the value of x.

x	–3	–2	–1	0	1	2
y	4	1	0	1	4	9

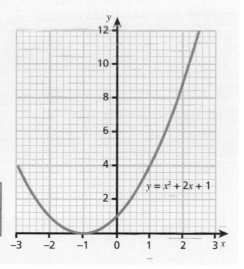

> **Quick Test**
>
> 1. What are the gradient and intercept of these equations?
>   a) $y = 3x + 5$ b) $y = 6x - 7$ c) $y = -3x + 2$
> 2. Fill in the table for the coordinates of $y = x^2 + 3x + 4$
>
> | x | –3 | –2 | –1 | 0 | 1 | 2 | 3 |
> |---|---|---|---|---|---|---|---|
> | y | | | | | | | |

> **Key Words**
>
> simultaneous equations
> quadratic equation
> substitute

Review Questions

3D Shapes: Volume and Surface Area

① Name the shape that has five faces, five vertices and eight edges. [1]

② Draw the nets of these shapes. [3]

a)

b)

c)

Total Marks _____ / 4

① Work out the surface area and the volume of this cuboid. Do not forget the units. [4]

14 cm

2 cm

6 cm

② Work out the volume of this oil drum and state the units. [3]

2.2 m

11 m

(MR) ③ If the volume of a cube is 512m³, what is the length of the sides in cm? [2]

(MR) ④ Phil has 1100cm² of paper to wrap this birthday present. [2]

Has he got enough paper?

20 cm

10 cm

15 cm

Total Marks _____ / 11

Interpreting Data

1 Lalana works at a call centre. The table shows the calls she took in one day.

Call length (*t* minutes)	Frequency
$0 < t \leqslant 2$	25
$2 < t \leqslant 4$	40
$4 < t \leqslant 6$	18
$6 < t \leqslant 8$	10
$8 < t \leqslant 10$	4

Draw a frequency diagram for the data. [3]

Total Marks _____ / 3

(MR) 1 Plot the following data on a scatter graph. Discuss any patterns and explain what the graph shows. [4]

TV viewing figures (in 1000s)	50	45	25	65	80	75	40	30	55
TV advert spend (in £1000s)	40	30	10	45	60	70	35	15	30

2 For each survey question below, state two things that could be improved.

a) Do you eat a lot of junk food? Yes ☐ No ☐

b) How much fruit do you eat in a week? 1 ☐ 2 ☐ 3 ☐ 4 ☐ [4]

Total Marks _____ / 8

Fractions

1 Prove, using the grids below, that $\frac{3}{5}$ is smaller than $\frac{2}{3}$.

[3]

2 Solve and simplify.

a) $\frac{4}{10} + \frac{1}{4} + \frac{2}{5} =$

b) $\frac{2}{5} + \frac{1}{8} + \frac{1}{2} =$

c) $\frac{3}{4} + \frac{2}{5} + \frac{3}{10} =$

d) $\frac{3}{4} - \frac{1}{8} =$

e) $\frac{5}{6} - \frac{1}{5} - \frac{1}{3} =$

f) $\frac{7}{9} - \frac{1}{4} =$ [6]

3 Solve and fully simplify.

a) $\frac{1}{8} \times \frac{2}{3} =$

b) $\frac{5}{6} \times \frac{8}{9} =$

c) $\frac{3}{10} \times \frac{1}{2} =$ [3]

4 Solve and fully simplify.

a) $\frac{1}{8} \div \frac{2}{3} =$

b) $\frac{1}{6} \div \frac{8}{9} =$

c) $\frac{3}{4} \div \frac{3}{7} =$ [3]

Total Marks _____ / 15

1 Solve each of these, giving your answer as a mixed number.

a) $4\frac{3}{8} + 2\frac{1}{5} =$ [2]

b) $3\frac{3}{5} + 2\frac{3}{9} =$ [2]

c) $7\frac{1}{4} - 2\frac{8}{11} =$ [2]

d) $2\frac{1}{5} - 1\frac{3}{7} =$ [2]

Total Marks _____ / 8

Coordinates and Graphs

1 The two points on this grid are the opposite corners of a square. Find the coordinates of the missing vertices (corners).

[2]

2 Draw a graph that shows the lines $x = -3$ and $y = 2$. [2]

3 Complete the table for the equation $y = 3x - 4$ and plot your results on a graph.

[3]

x	1	0	1	2	3	4
y						

Total Marks .... / 7

1 Complete the table below for the quadratic equation $y = x^2 + 3x - 2$

[3]

x	−3	−2	−1	0	1	2	3
y							

Total Marks .... / 3

Angles 1

You must be able to:

- Measure and draw angles
- Use properties in triangles to solve angle problems
- Use properties in quadrilaterals to solve angle problems
- Bisect an angle.

How to Measure and Draw an Angle

- **Angles** are measured in **degrees**, denoted by a little circle after the number, i.e. 45°.
- The piece of equipment used to measure and draw an angle is called a **protractor**.

> **Key Point**
>
> There are two scales on your protractor. This is so the protractor can be used in two different directions.
>
> Always start at zero and count up.

To measure the angle of a line:
- Place the protractor with the zero line on the base line.
- The centre should be level with the point where the two lines cross.
- Counting up from zero, count the degrees of the angle you are measuring.

To draw an angle:
- Draw a base line for the angle.
- Line up your protractor, putting the centre on the end of the line.
- Count up from zero until you reach your angle, e.g. 45°.
- Put a mark. Remove the protractor and draw a straight line joining the end of the base line to your point.

Angles in a Triangle

- Angles in any **triangle** add up to 180°. You can use this fact to help you work out missing angles.
- An isosceles triangle has two sides the same and two (base) angles the same.
- A right-angled triangle has a 90° angle, so if you have one more angle you can work out the remaining one.

Example

Find the missing angle x in these triangles.

$180° - (90° + 35°) = x° = 55°$

$180° - (53° \times 2) = x° = 74°$

Angles in a Quadrilateral

- Angles in any **quadrilateral** add up to 360°.

Example

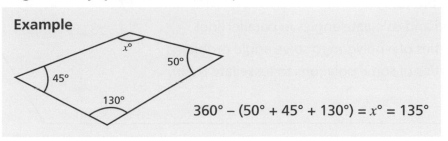

$$360° - (50° + 45° + 130°) = x° = 135°$$

- A parallelogram has two sets of equal angles. The opposite angles are equal.
- Only one angle is needed to be able to work out the others.

Example

$$360° - (110° \times 2) = 140°$$
Now share this angle equally between the remaining two corners:
$$x = 140° \div 2 = 70°$$

Bisecting an Angle

- **Bisect** means to cut exactly into two.

Example
1. Open your compass to a distance that is at least halfway along one of the lines.
2. Put the point of the compass where the two lines touch. Mark an arc on both lines.
3. Now place the point on one of these arcs and draw another arc between the lines.
4. Repeat for the arc on the other line, creating a cross.
5. Draw a line from where the lines touch through the cross.

Quick Test

1. Using a protractor draw an angle of:
   a) 48° b) 84° c) 125° d) 167°
2. Find the missing angle in each of these shapes.
   a) b) c)

Angles 2

You must be able to:

- Understand and calculate angles in parallel lines
- Use properties of a polygon to solve angle problems
- Use properties of some polygons to tessellate them.

Angles in Parallel Lines

- **Parallel** lines are lines that run at the same angle.
- Using parallel lines and a line that crosses them, you can apply some observations to help find missing angles.
- The angles represented by are equal. They are called **corresponding** angles.
- The angles represented by ♥ are also equal. They are called **alternate** angles.

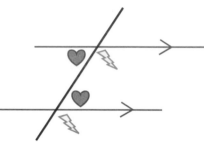

- Alternate angles are also called Z angles, because the lines look like a Z.
- Corresponding angles are also called F angles, because the lines look like an F.

Angles in Polygons

- A regular polygon is a shape that has each of its sides and angles equal.
- Using the fact that angles in a triangle add up to 180°, you can split any shape into triangles to help you work out the number of degrees inside that shape.

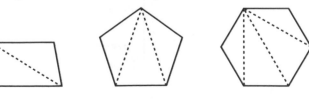

> **Key Point**
>
> A polygon can be split into a number of triangles. Try this formula to speed up calculation:
>
> (No. of sides – 2) × 180 = sum of interior angles

- Dividing the total number of degrees inside a regular polygon by the number of vertices will give the size of one **interior angle**.
- The sum of the **exterior angles** of any shape always equals 360°.

Shape	Number of sides	Sum of the interior angles
Triangle	3	180°
Quadrilateral	4	360°
Pentagon	5	540°
Hexagon	6	720°
Heptagon	7	900°
Octagon	8	1080°

Example

A regular polygon has interior angles of 140°. How many sides does it have?

The exterior angle of each part of the polygon is
180° – 140° = 40°

360° ÷ 40° = 9, so the polygon has nine sides.

Polygons and Tessellation

- Tessellation is where you repeat a shape or a number of shapes so they fit without any overlaps or gaps.

Key Point

You may notice that shapes that tessellate have angles that add up to 360° about one point.

Quick Test

1. Find the missing angles:

a)   b)   c)

2. What is the sum of the interior angles of a decagon?
3. Name a regular shape that would tessellate.

Key Words

parallel
corresponding
alternate
regular polygon
interior angle
exterior angle
tessellation

Probability 1

You must be able to:

- Recognise and use words associated with probability
- Construct and use a probability scale
- Calculate the probability of an event not occurring
- Construct and use sample spaces.

Probability Words

- Certain words are used to describe the **probability** of an **event** happening. How would you describe:
  - The probability of there being 40 days in a month?
    Impossible – there are at most 31 days in a month.
  - The probability that a student will attend school tomorrow?
    Likely – it cannot be said to be certain as the student might be on school holiday or ill and not attending school.
  - The probability that I roll a 2 on a dice?
    Unlikely – there are six possible outcomes and the number 2 is only one of these.
  - The probability that out of a bag with only green sweets in, a green sweet is pulled out?
    Certain – in this case there is no other outcome possible.
  - The probability a coin is tossed and it lands on heads?
    An **even chance** – the event outcome could be a head or a tail, two equally likely options.

Key Point

Try to consider the event with all the possible outcomes.

Once all the possible outcomes have been considered, the word can be selected.

Probability Scale

- A **probability scale** is a way of representing the probability words visually – putting the words we used above on a number line.
- The two extremes go at either end of the scale and the other probability words can be slotted in between:

```
        unlikely          likely
  |----------+--------------+----------|
impossible      even chance      certain
```

Example
Put an arrow on the probability scale above to represent the probability of pulling a red sweet out of a bag that contains 7 red sweets and 1 blue sweet.

- If an event has a more than equal outcome, then it is said to be **biased**.

Probability of an Event Not Occurring

- Probability can be thought of in terms of numbers as well as words. You could think of something happening as a fraction or a decimal.
- The probability of an event not happening can be summarised as 1 – event occurring.

Example

A non-biased six-sided dice is rolled:

P(rolling a 1) = $\frac{1}{6}$

P(rolling a 2) = $\frac{1}{6}$

P(rolling a 3) = $\frac{1}{6}$

P(rolling a 4) = $\frac{1}{6}$

P(rolling a 5) = $\frac{1}{6}$

P(rolling a 6) = $\frac{1}{6}$

P(not rolling a 4 or 5) = $1 - \left(\frac{1}{6} + \frac{1}{6}\right) = \frac{4}{6} = \frac{2}{3}$

Example

The probability of the weather being cloudy = 0.4

So the probability of it *not* being cloudy is 1 – 0.4 = 0.6

Sample Spaces

- A **sample space** shows all the possible outcomes of the event.

Example

The sample space for flipping a coin and rolling a dice is:

H1	H2	H3	H4	H5	H6
T1	T2	T3	T4	T5	T6

This shows all the possible outcomes and helps us to calculate the probability of events occurring.

Quick Test

1. How can the probability of rain in Manchester in October be described?
2. Show the event in question 1 on a probability scale.
3. A bag contains 5 red sweets and 10 blue sweets.
   a) What is the probability of picking a red sweet?
   b) What is the probability of not picking a red sweet?
4. If it rains 0.75 of the time in the rainforest and is cloudy 0.1 of the time, what is the probability it isn't raining or cloudy?

Key Words

probability
event
impossible
likely
unlikely
certain
even chance
probability scale
biased
sample space

Probability 2

You must be able to:

- Understand what mutually exclusive events are
- Calculate a probability with and without a table
- Work with experimental probability
- Understand Venn diagrams and set notation.

Mutually Exclusive Events

- **Mutually exclusive** events can be defined as things that can't happen at the same time.
- For example, this spinner can't land on yellow and blue at the same time as it only has one pointer. You can say that the probability of getting a yellow and a blue is 0, i.e. mutually exclusive.
- You could calculate the probability of spinning a yellow or a blue, which would be $\frac{1}{4} + \frac{1}{4} = \frac{1}{2}$
- Wearing one orange sock and one purple sock would not be mutually exclusive as these events could happen at the same time.

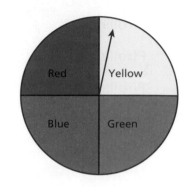

Calculating Probabilities and Tabulating Events

- Probability can be calculated by evaluating **all** outcomes.

Examples

What is the probability of picking a red ball from the bag?

The number of red balls
$\frac{7}{15}$
The total number of balls

What is the probability of picking out a green *or* a blue ball?

$\frac{3}{15} + \frac{4}{15} = \frac{7}{15}$

There are three different colours in this bag of 15 balls: 3 yellow, some red and some blue balls. Use the table to work out the number of each colour ball.

Yellow	Red	Blue
0.2	0.6	?

The probability that any ball is chosen is 1 so we can calculate the probability of drawing a blue ball: $1 - 0.6 - 0.2 = 0.2$

One way of working out the actual number of each colour is to multiply the probability by the total number of balls:
$15 \times 0.6 = 9$ (red) $0.2 \times 15 = 3$ (blue)

Key Point

Using the word **or** in probability usually implies that we add the probabilities of the events involved in the statement together.

Using the word **and** in probability can imply that we multiply the probabilities of the events.

Experimental Probability

- The probability of getting a 6 when rolling a six-sided dice is $\frac{1}{6}$. If you rolled the dice six times would you definitely get a 6? You may not get a 6 in six rolls, however the more times you roll the dice the more likely you are to get closer to $\frac{1}{6}$. This is **experimental probability**.

> **Key Point**
>
> You are using probabilities to guess events. These don't map exact outcomes but should be used to guide you to likely and less likely events.

Example

Hayley sat outside her school and counted 25 cars that went past. She noted the colour of each car in this table.

Yellow	1
Red	6
Blue	4
Black	9
White	5

a) What is the probability of the next car going past being white? $\frac{5}{25} = \frac{1}{5} = 0.2$

b) How many black cars would you expect if 50 cars go past?
$\frac{9}{25} = 0.36$ $0.36 \times 50 = 18$

Venn Diagrams and Set Notation

- Venn diagrams can be used to organise sets and find probabilities.

Example

Set A is the children who like green beans. Set B is the children who like carrots.

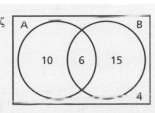

(A) = 10 + 6 = 16 children like green beans

(B) = 6 + 15 = 21 children like carrots

(A ∪ B) = 10 + 6 + 15 = 31 children like at least one. ← This is the union of A and B.

(A ∩ B) = 6 children like both green beans and carrots. ← This is the intersection of A and B.

(A ∪ B)' = 4 children like neither green beans nor carrots. ← The symbol ' means not in A or B.

> **Quick Test**
>
> 1. A bag has 20 balls of four different colours.
>   a) Complete the table.
>
> | Yellow | ? |
> |--------|-----|
> | Red | 0.3 |
> | Green | 0.2 |
> | Blue | 0.1 |
>
>   b) What is the probability of not getting a yellow ball?
>   c) If I picked out and replaced a ball 60 times, what is the expected probability that I would pick out a green ball?

> **Key Words**
>
> mutually exclusive
> experimental probability

Review Questions

Fractions

1 What fraction of each shape is shaded in? Give two other equivalent fractions for each. 📱 [3]

a) 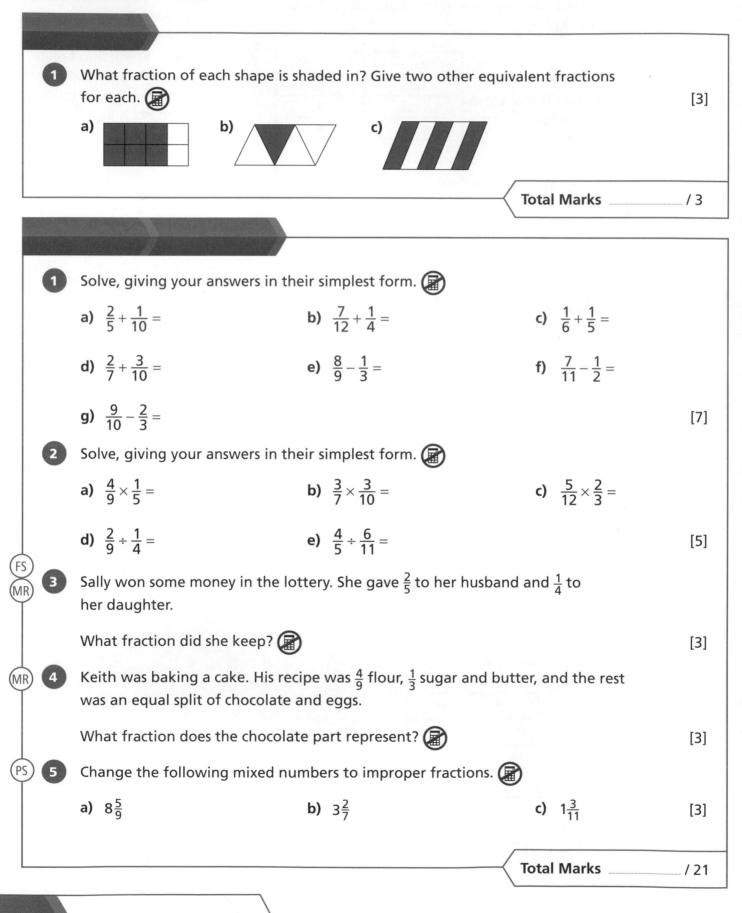   b)   c)

1 Solve, giving your answers in their simplest form. 📱

a) $\frac{2}{5} + \frac{1}{10} =$

b) $\frac{7}{12} + \frac{1}{4} =$

c) $\frac{1}{6} + \frac{1}{5} =$

d) $\frac{2}{7} + \frac{3}{10} =$

e) $\frac{8}{9} - \frac{1}{3} =$

f) $\frac{7}{11} - \frac{1}{2} =$

g) $\frac{9}{10} - \frac{2}{3} =$

[7]

2 Solve, giving your answers in their simplest form. 📱

a) $\frac{4}{9} \times \frac{1}{5} =$

b) $\frac{3}{7} \times \frac{3}{10} =$

c) $\frac{5}{12} \times \frac{2}{3} =$

d) $\frac{2}{9} \div \frac{1}{4} =$

e) $\frac{4}{5} \div \frac{6}{11} =$

[5]

FS
MR
3 Sally won some money in the lottery. She gave $\frac{2}{5}$ to her husband and $\frac{1}{4}$ to her daughter.

What fraction did she keep? 📱 [3]

MR **4** Keith was baking a cake. His recipe was $\frac{4}{9}$ flour, $\frac{1}{3}$ sugar and butter, and the rest was an equal split of chocolate and eggs.

What fraction does the chocolate part represent? 📱 [3]

PS **5** Change the following mixed numbers to improper fractions. 📱

a) $8\frac{5}{9}$

b) $3\frac{2}{7}$

c) $1\frac{3}{11}$

[3]

Coordinates and Graphs

1 Copy the grid shown and plot the following points:

a) (1, 2)   [1]

b) (−4, 5)   [1]

c) (3, −2)   [1]

2 On the same grid plot the following lines:

a) $x = 4$ **b)** $y = -1$ [2]

Total Marks .... / 5

1 Copy and complete the arrows and the table below for the equation $y = 2x - 4$. Then plot your results on a grid like the one shown.

| $\times$ | | — |

x	−1	0	1	2	3
y					

[4]

2 Copy and complete the table using the equation below.

$y = x^2 + 5x + 1$

x	−3	−2	−1	0	1	2	3
y							

[3]

Total Marks .... / 7

Angles

1 Find the missing angles in the following diagrams.

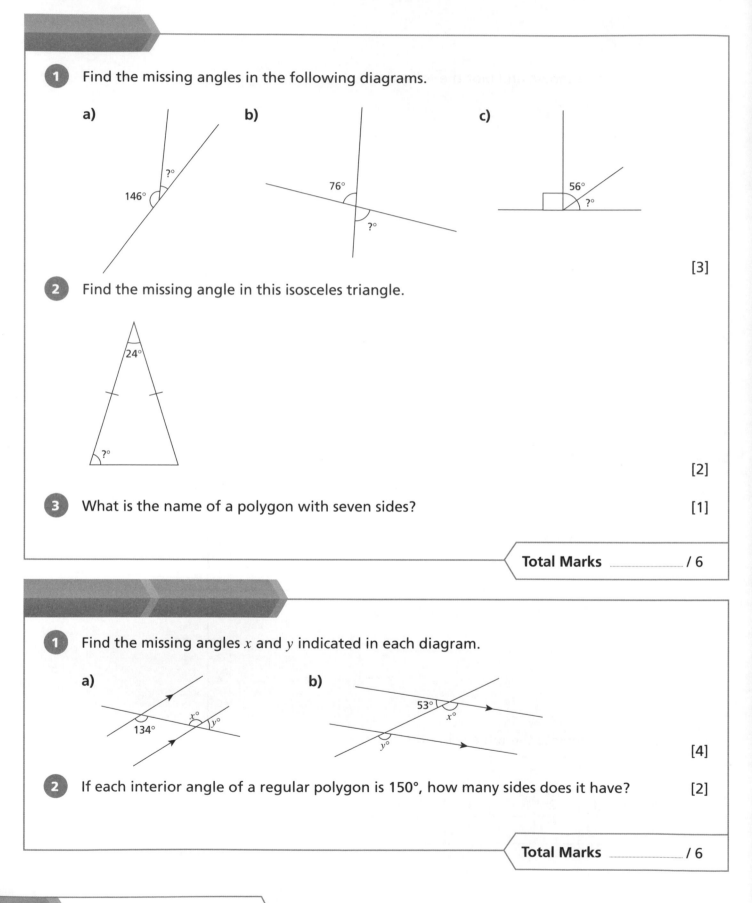

a)

b)

c)

[3]

2 Find the missing angle in this isosceles triangle.

[2]

3 What is the name of a polygon with seven sides? [1]

Total Marks .... / 6

1 Find the missing angles x and y indicated in each diagram.

a)

b)

[4]

2 If each interior angle of a regular polygon is 150°, how many sides does it have? [2]

Total Marks .... / 6

Probability

1 If today is Sunday, what is the chance that tomorrow is Tuesday? [1]

2 An eight-sided spinner is numbered 0, 0, 0, 1, 1, 1, 1, 2.

   a) What is the probability of getting a 0 on one spin? [1]

   b) What is the probability of **not** getting a 2? [2]

Total Marks _____ / 4

1 Phil does an experiment dropping a paper cup a number of times. The probability of it landing upside-down is 0.65

   What is the probability of it **not** landing upside-down? [2]

2 A dice has been rolled 50 times and recorded in the frequency table below.

   a) Copy and complete the table. Give the probabilities as fractions.

Number	Frequency	Estimated probability
1	5	
2	8	
3	7	
4	7	
5	8	
6	15	
Total 50		1

[3]

   b) Use the results in your table to work out the estimated probability (as fractions) of getting:

   i) the number 6   ii) an odd number   iii) a number bigger than 4 [3]

Total Marks _____ / 8

Fractions, Decimals and Percentages 1

You must be able to:

- Convert between a fraction, decimal and percentage
- Calculate a fraction of a quantity
- Calculate a percentage of a quantity
- Compare quantities using percentages.

Different Ways of Saying the Same Thing

- The table below shows how **fractions**, **decimals** and **percentages** are used to represent the same thing:

Picture	Fraction	Decimal	Percentage
$\frac{1}{4}$	$\frac{1}{4}$	0.25	25%
$\frac{1}{2}$	$\frac{1}{2}$	0.5	50%
$\frac{3}{4}$	$\frac{3}{4}$	0.75	75%

Key Point

Percent means 'out of 100'.

Converting Fractions to Decimals to Percentages

- For conversions you do not know automatically, use the rules below.

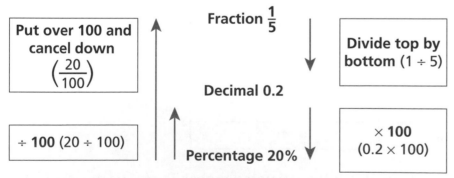

Fraction $\frac{1}{5}$

Put over 100 and cancel down $\left(\frac{20}{100}\right)$

Divide top by bottom (1 ÷ 5)

Decimal 0.2

÷ **100** (20 ÷ 100)

× **100** (0.2 × 100)

Percentage 20%

Fractions of a Quantity

- To find a fraction of a **quantity** without a calculator, divide by the **denominator** and multiply by the **numerator**.

Example

Find $\frac{2}{3}$ of £120.

$$= 120 \div 3 \times 2 = £80$$

- To find a fraction of a quantity with a calculator, use the fraction button . Calculators can differ though, so find out how yours deals with fractions.

Example
Find $\frac{2}{5}$ of £200.

$= 2 \; \blacksquare \; 5 \times 200 = £80$

Percentages of a Quantity

- To find a percentage of a quantity without a calculator:

Example
Find 20% of £60.

$$10\% \text{ of } £60 = 60 \div 10 = £6$$
$$20\% = £6 \times 2 = £12$$

← 20% is two lots of 10%

- To find a percentage of a quantity with a calculator:

Example
Find 20% of £60.

$= 20\% \times £60$
$= 20 \div 100 \times 60 = £12$

Or 20% = 0.2
so 20% of £60 is
$0.2 \times £60 = £12$

← Change the percentage to a decimal.

Example
Find 120% of £60.

$= 120\% \times £60$
$= 120 \div 100 \times 60 = £72$

Or 120% = 1.2
so 120% of £60 is
$1.2 \times £60 = £72$

← Remember: as 120% > 100%, your answer will be bigger than £60.

- To compare two quantities using percentages:

Example
Which is bigger, 30% of £600 or 25% of £700?

10% of £600 = 600 ÷ 10 = £60   50% of £700 = 700 ÷ 2 = £350
30% of £600 = 60 × 3 = £180   25% of £700 = 350 ÷ 2 = £175

So 30% of £600 is bigger, by £5.

> **Key Point**
>
> Useful percentages to know:
>
> 50% → ÷ 2
>
> 10% → ÷ 10
>
> 1% → ÷ 100
>
> 'of' means '×'

> **Key Words**
>
> **fraction**
> **decimal**
> **percentage**
> **quantity**
> **denominator**
> **numerator**

> **Quick Test**
>
> 1. Change $\frac{7}{20}$ to a decimal and a percentage.
> 2. Change 36% to a fraction in its lowest terms.
> 3. Work out $\frac{2}{5}$ of $70.
> 4. Find 35% of $140.

Fractions, Decimals and Percentages 2

You must be able to:

- Work out a percentage increase or decrease
- Find one quantity as a percentage of another
- Work out simple interest.

Percentage Increase and Decrease

- To find a percentage **increase** or **decrease**, add on or subtract the percentage you have found.

Example

A calculator is priced at £12 but there is a discount of 25%.
Work out the reduced price of the calculator.

$$25\% \text{ of } £12$$
$$= 25\% \times £12$$
$$= 25 \div 100 \times 12 = £3$$

Reduced price = £12 − £3 = £9

A one-step way:

A reduction of 25% means you are left with 75%, and 75% = 0.75
0.75 × £12 = £9

£3 is the discount so **'take it away'** to get the final answer.

Example

A laptop computer costs £350 plus tax at 20%.
Work out the actual cost of the laptop.

$$20\% \text{ of } £350$$
$$= 20\% \times £350$$
$$= 20 \div 100 \times 350 = £70$$

Actual cost = £350 + £70 = £420

A one-step way:

An increase of 20% means you pay 120%, and 120% = 1.2
1.2 × £350 = £420

£70 is the tax so **'add it on'** to get the final answer.

Finding One Quantity as a Percentage of Another

- To find one quantity as a percentage of another:

Example

Jane gets 18 out of 20 in a test. What percentage is this?

With a calculator:

$$\frac{18}{20} \times 100$$
$$= 18 \div 20 \times 100$$
$$= 90\%$$

Change the fraction to a decimal.

Without a calculator:

$$\frac{18}{20} = \frac{90}{100} = 90\%$$

Make the fraction 'out of' 100.

Simple Interest

- Find the **interest** for one year then multiply by the number of years.

Example

Peter puts £200 into a savings account. He gets 5% simple interest per year.

How much does he have in his account after two years?

10% of £200
= 200 ÷ 10
= £20

5% = 20 ÷ 2 = £10

After two years Peter receives £10 × 2 = £20

Peter has £200 + £20 = £220 in his account.

Key Point

Interest is **not** added on at the end of each year.

This is the interest for one year.

Example

Sarah borrows £2000 from her bank. She has to pay back the loan with additional interest at 8% per year.

After three years, how much will Sarah have to pay back?

8% of £2000
= 8% × £2000
= 8 ÷ 100 × 2000 = £160

After three years the interest will be £160 × 3 = £480

So Sarah will have to pay back £2000 + £480 = £2480

This is the interest for one year.

Quick Test

1. A television costing £450 is reduced by 10%. What is its sale price?
2. A house costing £80000 increases in value by 15%. What is the cost of the house now?
3. Anne gets $\frac{21}{25}$ in a test. What percentage is this?

Key Words

increase
decrease
interest

Equations 1

You must be able to:

- Find an unknown number
- Solve a simple equation
- Solve an equation with unknowns on both sides
- Apply the inverse of an operation.

Finding Unknown Numbers

- The unknown number is usually given as a letter or symbol.
- Applying the **inverse** operations helps you to find the unknown number.

Example

$n + 4 = 13$ or $\blacksquare + 4 = 13$

Take 4 away from 13.

n or $\blacksquare = 13 - 4$

n or $\blacksquare = 9$ (check: $9 + 4 = 13$)

← Both ways mean **'something + 4 = 13'**.

← -4 is the 'inverse' or 'opposite' of $+4$.

Example

$x - 4 = 13$ or $\bigcirc - 4 = 13$

Add 4 to 13.

x or $\bigcirc = 13 + 4$

x or $\bigcirc = 17$ (check: $17 - 4 = 13$)

← Both ways mean **'something − 4 = 13'**.

← $+4$ is the 'inverse' or 'opposite' of -4.

- The same idea applies to multiplying and dividing.

Example

$3n = 12$

$n = 12 \div 3$

$n = 4$ (check: $3 \times 4 = 12$)

← $3 \times$ something $= 12$

← $\div 3$ is the inverse of $\times 3$

Example

$\frac{n}{3} = 4$

$n = 4 \times 3$

$n = 12$ (check: $12 \div 3 = 4$)

← Something $\div 3 = 4$

← $\times 3$ is the inverse of $\div 3$

- Now you can apply inverse operations to more difficult **equations**.

Key Point

Operation	Inverse
$+$	$-$
$-$	$+$
$\times$	$\div$
$\div$	$\times$

Solving Equations

- Remember to think of the letter as 'something'.

Example

Solve the equation $2y + 3 = 15$

This simply means 'something' $+ 3 = 15$

So $2y = 12$ means $2 \times$ 'something' $= 12$

So $y = 6$

Now look with the **inverses**:

$$2y + 3 = 15$$

$(- 3)$ $\quad 2y = 12$

$(\div 2)$ $\quad y = 6$

← 'Something' must be 12.

← 'Something' must be 6.

> **Key Point**
>
> Remember to do the same to both sides of the equation.

Equations with Unknowns on Both Sides

- An equation may have an unknown number on both sides of the equals sign.

Example

Solve the equation $5x - 2 = 3x + 5$

$$5x - 2 = 3x + 5$$

$(- 3x)$ $\quad 2x - 2 = 5$

$(+ 2)$ $\quad\quad 2x = 7$

$(\div 2)$ $\quad\quad x = 3.5$

← Do the same to both sides.

← Get rid of all the xs on one side of the equation.

← Now do the inverses.

Example

Solve the equation $3x + 5 = 5x - 4$

$$3x + 5 = 5x - 4$$

$(- 3x)$ $\quad 5 = 2x - 4$

$(+ 4)$ $\quad 9 = 2x$

$(\div 2)$ $\quad 4.5 = x$

← This is the same as $x = 4.5$

> **Quick Test**
>
> 1. What is the inverse of $\times 6$?
> 2. If $\bigcirc - 5 = 7$, what is the value of $\bigcirc$?
> 3. If $6n = 30$, what is the value of n?
> 4. Solve the equation $3y - 2 = 13$
> 5. Solve the equation $3x + 7 = 2x - 2$

> **Key Words**
>
> inverse
> equation
> solve

Equations 2

You must be able to:

- Solve more complex equations with fractions or negative numbers
- Set up and solve an equation
- Apply the inverse of an operation
- Use negative numbers.

Solving More Complex Equations

- An equation may include a **negative** of the unknown number.
- The unknown number may be part of a fraction or inside **brackets**.

Example

Solve the equation $3x + 1 = 11 - 2x$

$$3x + 1 = 11 - 2x$$
$(+ 2x)$ $\quad 5x + 1 = 11$
$(- 1)$ $\qquad 5x = 10$
$(\div 5)$ $\qquad x = 2$

> Be careful. This equation has some **negative** values of x. Adding $2x$ to both sides gets rid of the negative xs.

Example

Solve the equation $\dfrac{3x + 1}{2} = 8$

$$\dfrac{3x + 1}{2} = 8$$
$(\times 2)$ $\quad 3x + 1 = 16$
$(- 1)$ $\qquad 3x = 15$
$(\div 3)$ $\qquad x = 5$

> Here the $\times 2$ cancels out the $\div 2$ on the left-hand side of the equation.

Example

Solve the equation $3(2x - 1) = 4(x + 2)$

Multiply out the brackets first then solve in the usual way:

$$3(2x - 1) = 4(x + 2)$$
$$6x - 3 = 4x + 8$$
$(- 4x)$ $\quad 2x - 3 = 8$
$(+ 3)$ $\qquad 2x = 11$
$(\div 2)$ $\qquad x = 5.5$

Key Point

Remember to multiply everything inside the bracket by the number outside the bracket.

Setting Up and Solving Equations

- You will have to follow a set of instructions in a given order.
- Usually you only have to ×, ÷, +, − or square.
- If you have to multiply 'everything' then remember to use brackets.

Example

A number is **doubled**, then 5 is added to the total and the result is 11.

What was the original number?

The words	The algebra
A number	n
doubled	$2n$
add 5	$2n + 5$
the result is 11	$2n + 5 = 11$

You can now solve this in the usual way to find the original number equals 3.

Example

Three boys were paid £10 per hour plus a tip of £6 to wash some cars. They shared the money and each got £12. How many hours did they wash cars for?

£10 × number of hours   £6 tip

Set up an equation: $\dfrac{10x + 6}{3} = 12$ each got

shared by 3 boys

Now solve the equation:

$$(\times 3) \quad 10x + 6 = 36$$
$$(-6) \quad\quad 10x = 30$$
$$(\div 10) \quad\quad x = 3$$

The boys worked for 3 hours.

Key Point

Remember to do the same to both sides of the equation.

Quick Test

1. Solve the equation $3x - 4 = 11$
2. Solve the equation $2x + 3 = 12 - x$
3. Solve the equation $2(x + 2) = 2(3x - 2)$
4. A number is multiplied by 3 and then 8 is subtracted. The result is 25. What is the number?
5. A chicken is roasted for 60 minutes for every kilogram, plus an extra 20 minutes. If the chicken took 140 minutes to cook, how heavy was it?

Key Words

negative
brackets
double

Review Questions

Angles

1 Find the missing angles.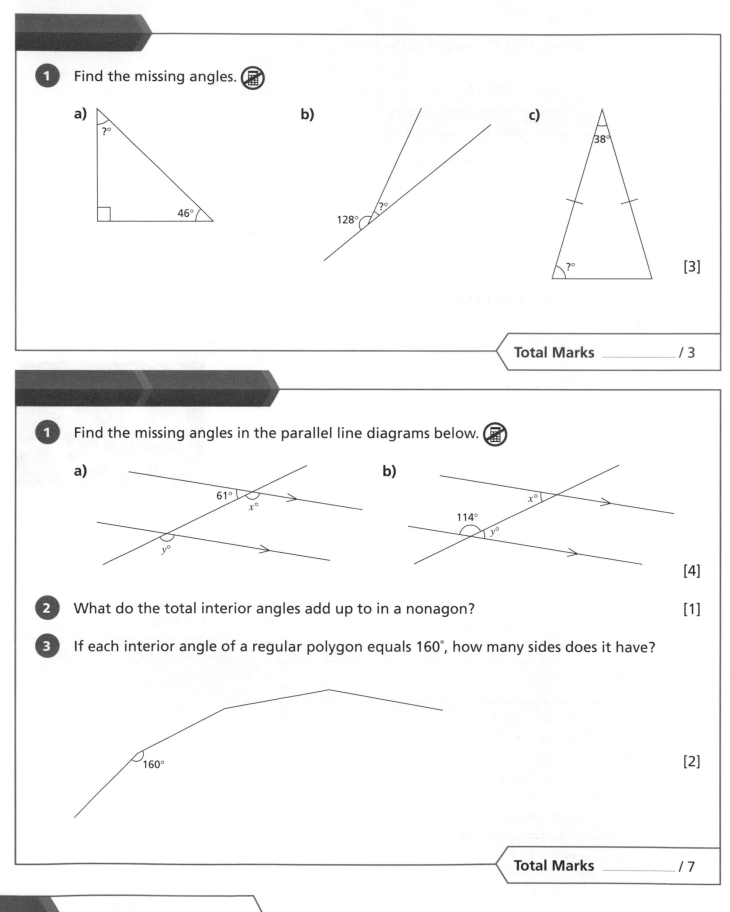

a)

?°

46°

b)

128°   ?°

c)

38°

?°

[3]

1 Find the missing angles in the parallel line diagrams below.

a)

61°
x°

y°

b)

x°

114°
y°

[4]

2 What do the total interior angles add up to in a nonagon?   [1]

3 If each interior angle of a regular polygon equals 160°, how many sides does it have?

160°

[2]

Probability

1 Copy the probability scale and complete the boxes with suitable probability words.

[2]

2 Ben has a drawer of 10 t-shirts: 3 blue, 4 yellow, 2 red and 1 pink. What is the probability that:

a) Ben pulls out a blue t-shirt? [1]

b) Ben pulls out either a yellow t-shirt or a red t-shirt? [1]

c) Ben does not pull out a pink t-shirt? [2]

Total Marks .... / 6

1 Leanne runs an ice-cream van. At random, she chooses which kind of sprinkles to put on the ice-creams. The table below shows what Leanne did on Sunday.

Sprinkles	Frequency	Probability
Chocolate	19	
Hundreds and thousands	14	
Strawberry	7	
Nuts	10	

a) Complete the experimental probabilities in the table above. [2]

b) What was the probability of getting either nuts or chocolate sprinkles? [2]

2 If the probability of winning a raffle prize is 0.47, what is the probability of not winning a raffle prize? [1]

3 a) Copy and complete the table below.

Sales destination	Probability of going to destination
London	0.26
Cardiff	0.15
Chester	0.2
Manchester	

[1]

b) Which is the least likely destination to travel to for sales? [1]

Total Marks .... / 7

Practice Questions

Fractions, Decimals and Percentages

1 Copy and complete the following table.

Fraction	Decimal	Percentage
$\frac{7}{10}$		
		55
	0.32	
$\frac{3}{100}$		

[4]

2 Work out: 📱

   a) 50% of £32   [1]

   b) 10% of 80cm   [1]

   c) 15% of 160m   [2]

   d) 25% of £104   [1]

3 Use the $a^{b}/_{c}$ button on your calculator to work out:

   a) $\frac{1}{5}$ of £85   [1]

   b) $\frac{2}{3}$ of £120   [1]

   c) $\frac{5}{7}$ of 21m   [1]

Total Marks _____ / 12

1 Work out the following, showing your working. 📱

   a) $\frac{2}{3}$ of £15   [3]

   b) $\frac{3}{7}$ of £210   [3]

   c) $\frac{4}{5}$ of £6000   [3]

PS 2 A jacket costing £75 is reduced by 20% in a sale. What is the sale price of the jacket? 📱 [3]

PS 3 In a Maths test, Karim scored 16 out of 20 and John scored 15 out of 20.

       What percentage did Karim and John each get?   [4]

Total Marks _____ / 16

Equations

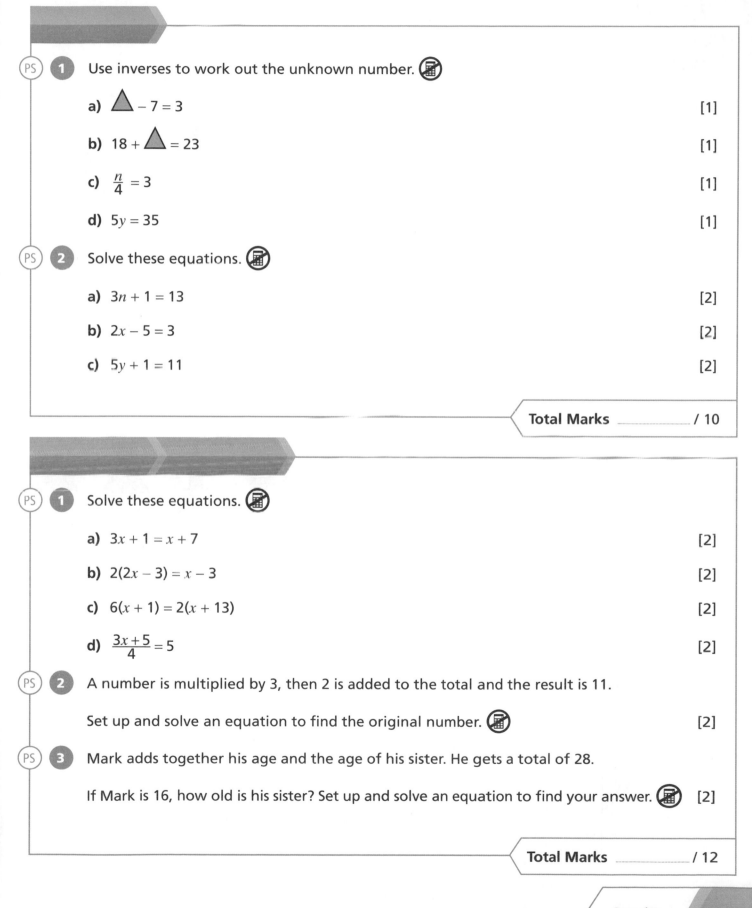

(PS) **1** Use inverses to work out the unknown number. 🔢

     a) ▲ − 7 = 3   [1]

     b) 18 + ▲ = 23   [1]

     c) $\frac{n}{4} = 3$ [1]

     d) $5y = 35$ [1]

(PS) **2** Solve these equations. 🔢

     a) $3n + 1 = 13$ [2]

     b) $2x - 5 = 3$ [2]

     c) $5y + 1 = 11$ [2]

Total Marks _____ / 10

(PS) **1** Solve these equations. 🔢

     a) $3x + 1 = x + 7$ [2]

     b) $2(2x - 3) = x - 3$ [2]

     c) $6(x + 1) = 2(x + 13)$ [2]

     d) $\frac{3x + 5}{4} = 5$ [2]

(PS) **2** A number is multiplied by 3, then 2 is added to the total and the result is 11.

     Set up and solve an equation to find the original number. 🔢   [2]

(PS) **3** Mark adds together his age and the age of his sister. He gets a total of 28.

     If Mark is 16, how old is his sister? Set up and solve an equation to find your answer. 🔢   [2]

Total Marks _____ / 12

Symmetry and Enlargement 1

You must be able to:

- Reflect a shape
- Translate a shape
- Find the order of rotational symmetry
- Rotate a shape
- Enlarge a shape.

Reflection and Reflectional Symmetry

- Reflect each point one at a time.
- Use a line that is **perpendicular** to the mirror line.
- Make sure the **reflection** is the same distance from the mirror line as the original shape.
- A shape has reflectional symmetry if you can draw a mirror line through it.

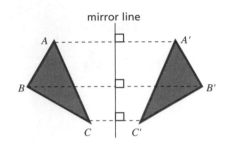

Translation

- Translation shifts a shape left/right (x) and/or up/down (y).
- The shift is given as a vector $\begin{pmatrix} x \\ y \end{pmatrix}$.

Rotational Symmetry

- A shape has rotational symmetry if it looks exactly like the original shape when it is **rotated**.
- The **order of rotational symmetry** is the number of ways the shape looks the same.
- To rotate a shape you need to know:
  - The **centre of rotation**
  - The direction it will rotate
  - The number of degrees to rotate it.

order 1   order 2   order 3   order 4

> **Key Point**
>
> Perpendicular means at right angles (90°) to.
>
> Check the position of your reflection by placing a mirror along the mirror line.

Example

a) Rotate shape A 180° from (1, 0).

b) Translate shape A by vector $\begin{pmatrix} -4 \\ 1 \end{pmatrix}$.

Enlargement

- To draw an **enlargement** you need to know two things:
  - How much bigger/smaller to make the shape. This is called the **scale factor**.
  - Where you will enlarge the shape from. This is called the **centre of enlargement**.
- Remember that enlarged shapes are **similar** (the same shape but a different size).

> **Key Point**
>
> If the scale factor is more than 1, the shape will be bigger.
>
> If the scale factor is less than 1, e.g. $\frac{1}{2}$, the shape will be smaller.
>
> Sometimes you will not be given a centre of enlargement and can do the drawing anywhere.

Example
Enlarge shape A by a scale factor of 2 from the point (3, 4).

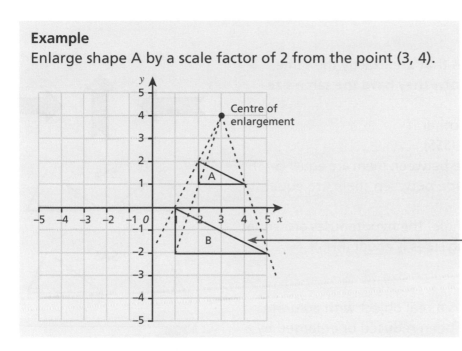

Enlarge every side of the shape.

- Use **rays** to check the position of your enlargement. They will touch corresponding corners of the shape.

Quick Test

1. What is the order of rotational symmetry of this shape?

2. a) Reflect this shape across the mirror line.
   b) Rotate the shape 180° about the corner A.

   c) Rotate the shape 90° clockwise about the corner B.
3. Enlarge this shape by a scale factor of 3.

1 cm
2 cm

> **Key Words**
>
> perpendicular
> reflection
> rotation
> centre of rotation
> enlargement
> scale factor
> centre of enlargement
> similar
> ray

Symmetry and Enlargement 2

You must be able to:

- Recognise congruent shapes
- Interpret a scale drawing
- Work out missing sides in similar shapes
- Convert between units of measure.

Congruence

- Congruence simply means shapes that are exactly the same.
- These arrow shapes are congruent – they have the same size and the same shape.
- Triangles are congruent to each other if:
  - three pairs of sides are equal (SSS)
  - two pairs of sides and the angle between them are equal (SAS)
  - two pairs of angles and the side between them are equal (ASA)
  - both triangles have a right angle, the hypotenuses are equal and one pair of corresponding sides is equal (RHS).

Same size   Same shape

SSS

SAS

Scale Drawings

- A scale drawing is one that shows a real object with accurate dimensions, except they have all been reduced or enlarged by a certain amount (called the scale).
- Similar shapes are enlarged by the same scale factor, but the angles stay the same.

> **Key Point**
>
> Remember, **all** sides must be multiplied by the same number.

Example

A scale of 1 : 10 means in the real world the object would be 10 times bigger than in the drawing. These horses are similar – same shape, different size.

- We use scale drawings to make copies of real objects.

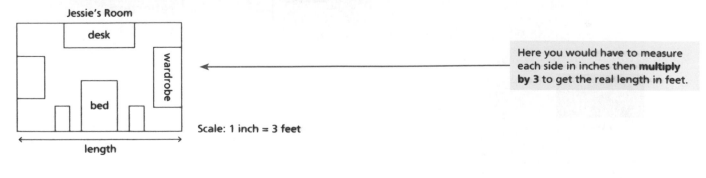

Jessie's Room

desk

wardrobe

bed

length

Scale: 1 inch = 3 feet

Here you would have to measure each side in inches then **multiply by 3** to get the real length in feet.

Shape and Ratio

- You can use **ratios** to work out the 'real' size of an object. The scale is given as a ratio with the smaller **unit** first.

Example
Estimate the height of this house using the scale of 1 : 160

5cm

6cm

For every 1cm you measure in the picture, multiply by 160 to get the real size. Then convert to metres.

> **Key Point**
>
> Scales are given as a ratio, usually 1 : n where n is what you multiply by.

Here, the height of the house is 5cm so in real life the house is actually:

$5 \times 160 = 800$cm $= 8$m

Remember: 100cm = 1m

The width of the house is 6cm so in real life the house is actually:

$6 \times 160 = 960$cm $= 9$m 60cm

> **Quick Test**
>
> 1. Draw three congruent shapes.
> 2. Which shape is congruent to shape A?
>
>
>
> A   B   C   D
>
> 3. Estimate the length, in metres, of the boat. Scale: 1 : 80
>
> 4. Which of these shapes are similar?
>
> A   B   C
>
> 2cm   4cm   2cm
>
> 3cm   6cm   6cm
>
> 5. Draw two shapes that are similar.

> **Key Words**
>
> congruent
> scale
> ratio
> units

Ratio and Proportion 1

You must be able to:

- Understand what ratio means
- Simplify a ratio
- Multiply and divide by whole numbers.

Introduction to Ratios

- Ratio is a way of showing the relationship between two numbers.
- Ratios can be used to compare costs, weights and sizes.

Example

On the deck of a boat there are 2 women and 1 man. There are also 5 cars and 2 bicycles.

The ratio of men to women is 1 to 2, written 1 : 2

The ratio of women to men is 2 to 1, written 2 : 1

The ratio of cars to bicycles is 5 to 2, written 5 : 2

The ratio of bicycles to cars is 2 to 5, written 2 : 5

Example

A recipe for making pastry uses 4oz flour and 2oz butter.

The ratio of flour to butter is 4 to 2, written 4 : 2

The ratio of butter to flour is 2 to 4, written 2 : 4

Example

What is the ratio of black tiles to blue tiles?

The ratio of black tiles to blue tiles is 5 : 9

- Ratios can also be written as fractions.
  For example, $\frac{2}{3}$ are women and $\frac{1}{3}$ are men.
  $\frac{5}{7}$ of the vehicles are cars and $\frac{2}{7}$ of the vehicles are bicycles.

Simplifying Ratios

- The following ratios are **equivalent**. The relationship between each pair of numbers is the same:

10 : 20
↓
3 : 6
↓
2 : 4
↓
1 : 2   This is a **simpler** way of writing the ratio 10 : 20

- You can **simplify** a ratio if you can divide by a common **factor**.
- When a ratio cannot be simplified, it is said to be in its **lowest terms**.

Example
Simplify the ratio 30 : 100

30 ÷ 10   3 : 10   100 ÷ 10   ← Divide both numbers by 10.

Example
Write 40 cents to $1 as a ratio in its lowest terms.

First get the units the same: 40 cents to 100 cents written

40 : 100
↓
Now simplify   (÷ 10)   4 : 10
↓
and again   (÷ 2)   2 : 5   ← This is now in its lowest terms.

Example
The angles of a triangle are 20°, 60° and 100°.

What is the ratio of the angles in its lowest terms?

20° : 60° : 100°

(÷ 20)   1 : 3 : 5   ← This is now in its lowest terms.

Quick Test

1. Look at this pattern of grey and green tiles:

   a) Write down the ratio of green tiles to grey tiles.
   b) Write down the ratio of grey tiles to green tiles.
2. Write the following ratios in their lowest terms:
   a) 3 : 9   b) 28 : 4   c) 25cm : 1m

Ratio and Proportion 2

You must be able to:

- Share in a given ratio
- Multiply and divide by whole numbers
- Work out proportional amounts
- Use the unitary method.

Sharing Ratios

- Sharing ratios are used when a total amount is to be **shared** or **divided** into a given ratio.

Example

Share £200 in the ratio 5 : 3

Add the ratio to find how many parts there are.

$$5 + 3 = 8 \text{ parts}$$

Divide £200 by 8 to find out how much 1 part is.

$$200 \div 8 = 25$$

$$1 \text{ part is } £25$$

Now multiply by each part of the ratio.

$$5 \times £25 = £125$$

$$3 \times £25 = £75$$

£200 shared in the ratio 5 : 3 is £125 : £75

Key Point

Divide to find one, then multiply to find all.

Example

A sum of money is shared in the ratio 2 : 3

If the smaller share is £30, how much is the sum of money?

2 parts = £30

so 1 part = £30 ÷ 2 = £15

3 parts = £15 × 3 = £45

The sum of money = £30 + £45 = £75

Direct Proportion

- Two quantities are in **direct proportion** if their ratios stay the same as the quantities get larger or smaller.

> **Example**
> If the ratio of teachers to students in one class is 1 : 30, then three classes will need 3 : 90

Using the Unitary Method

- Using the unitary method, find the value of **one unit** of the quantity before working out the required amount.

> **Example**
> Five loaves of bread cost £4.25. How much will three loaves cost?
>
> One loaf costs £4.25 ÷ 5 = 85 pence
>
> Three loaves will cost 85 pence × 3 = £2.55

Remember: divide to find one, then multiply to find all.

> **Example**
> This recipe for making apple pie serves four people:
>
> 200g flour
>
> 200g butter
>
> 50g sugar
>
> 8 large apples
>
> Change these amounts so there are enough ingredients for 10 people.
>
> Divide to find one, then multiply to find all.
>
> 200g flour ÷ 4 × 10   = 500g flour
>
> 200g butter ÷ 4 × 10   = 500g butter
>
> 50g sugar ÷ 4 × 10   = 125g sugar
>
> 8 large apples ÷ 4 × 10 = 20 large apples

Key Point

All the amounts have increased in proportion (by $2\frac{1}{2}$ times in this example).

> **Quick Test**
>
> 1. Share 40 sweets in the ratio 2 : 5 : 1
> 2. £360 is divided between Sara and John in the ratio 5 : 4 How much did each person receive?
> 3. Work out the missing ratio.
>   a) 3 : 5 = 12 : ?   b) 4 : 5 = ? : 35
> 4. If six CDs cost £30, how much will eight CDs cost?

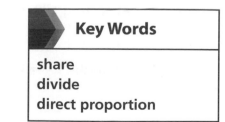

Key Words

share
divide
direct proportion

Review Questions

Fractions, Decimals and Percentages

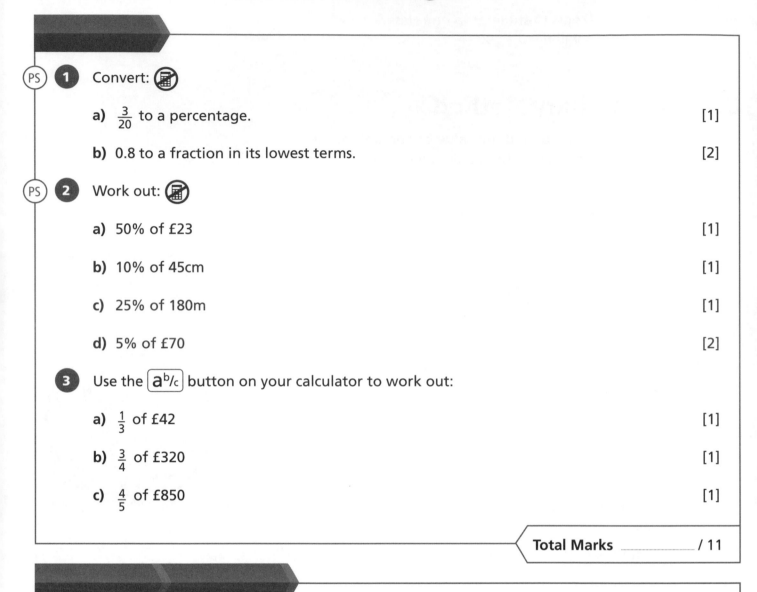

(PS) **1** Convert: 📱

    a) $\frac{3}{20}$ to a percentage. [1]

    b) 0.8 to a fraction in its lowest terms. [2]

(PS) **2** Work out: 📱

    a) 50% of £23 [1]

    b) 10% of 45cm [1]

    c) 25% of 180m [1]

    d) 5% of £70 [2]

3 Use the $a^b/_c$ button on your calculator to work out:

    a) $\frac{1}{3}$ of £42 [1]

    b) $\frac{3}{4}$ of £320 [1]

    c) $\frac{4}{5}$ of £850 [1]

Total Marks _____ / 11

(FS) **1** Jenny receives £5 pocket money every week.

    She spends $\frac{1}{2}$ of her money on magazines and $\frac{2}{5}$ on sweets. The rest she saves. 📱

    a) How much does Jenny spend on sweets? [2]

    b) How much does Jenny save? [2]

(PS) **2** A coat costing £90 is reduced by 15% in a sale. What is the sale price of the coat? 📱 [3]

(FS) **3** Kim puts £150 into a savings account. She will receive 6% simple interest each year. How much will she have in the bank after four years? [3]

Total Marks _____ / 10

Equations

(PS) **1** Use inverses to work out the unknown number. 📷

a) $7 \bigcirc = 28$ [1]

b) $n + 8 = 17$ [1]

c) $\frac{p}{3} = 6$ [1]

d) $\bigcirc - 7 = 12$ [1]

(PS) **2** Solve these equations. Show your working. 📷

a) $4n - 1 = 11$ [2]

b) $5x + 1 = 21$ [2]

c) $3a + 8 = 5$ [2]

Total Marks _____ / 10

(PS) **1** Solve these equations. Show your working. 📷

a) $6x - 5 = 4x + 7$ [2]

b) $5(x + 2) = 2(x - 1)$ [2]

c) $3x - 1 = 4 - 2x$ [2]

(FS) **2** Five builders are together paid £20 per hour plus a bonus of £150. They share the pay and each get £50.

Set up and solve an equation to find how many hours they worked. 📷 [2]

3 A chocolate bar machine holds 56 bars of chocolate.

If 29 are left, how many were sold? Set up and solve an equation to find your answer. 📷 [2]

Total Marks _____ / 10

Practice Questions

Symmetry and Enlargement

PS **1** Reflect shape A across the dotted mirror lines.

[3]

PS **2** What is the order of rotational symmetry of this shape?

[1]

Total Marks _____ / 4

PS **1** a) Rotate shape A 90° clockwise about the origin (0, 0). Label the new shape B. [2]

b) Enlarge shape A by a scale factor 3, with a centre of enlargement (3, 4). Label the new shape C. [2]

c) Which shapes are congruent? [1]

MR **2** A photograph 5cm × 7cm is to be enlarged by a scale factor of 4.

What are the dimensions of the new photograph? [2]

Total Marks _____ / 7

Ratio and Proportion

PS **1** What is the ratio of black tiles to white tiles? [1]

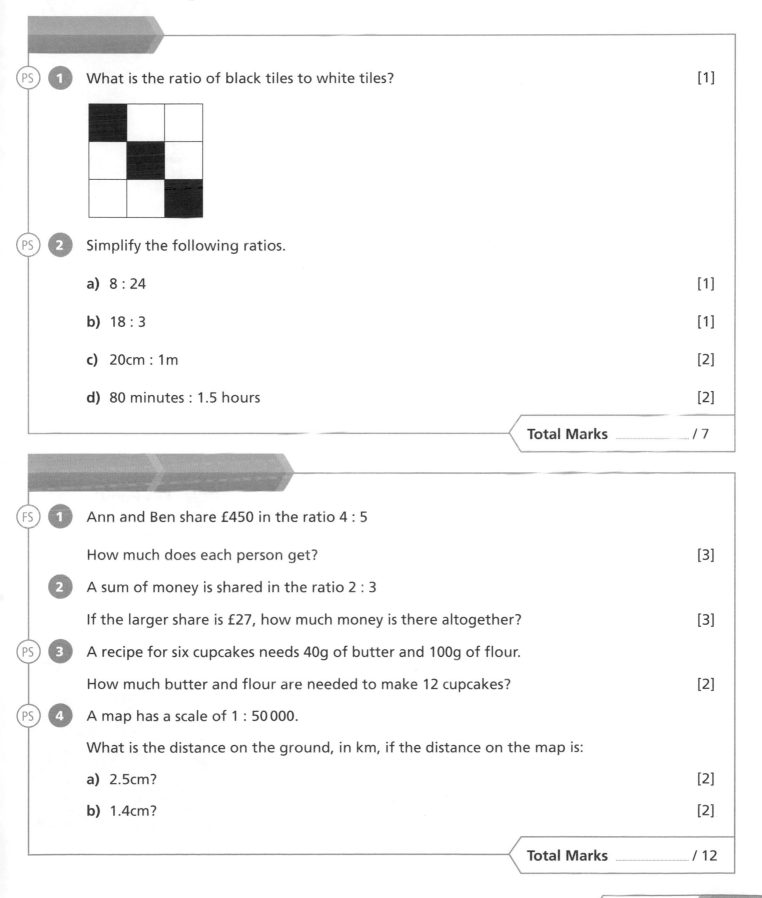

PS **2** Simplify the following ratios.

　a) 8 : 24 [1]

　b) 18 : 3 [1]

　c) 20cm : 1m [2]

　d) 80 minutes : 1.5 hours [2]

Total Marks _____ / 7

FS **1** Ann and Ben share £450 in the ratio 4 : 5

　How much does each person get? [3]

2 A sum of money is shared in the ratio 2 : 3

　If the larger share is £27, how much money is there altogether? [3]

PS **3** A recipe for six cupcakes needs 40g of butter and 100g of flour.

　How much butter and flour are needed to make 12 cupcakes? [2]

PS **4** A map has a scale of 1 : 50 000.

　What is the distance on the ground, in km, if the distance on the map is:

　a) 2.5cm? [2]

　b) 1.4cm? [2]

Total Marks _____ / 12

Real-Life Graphs and Rates 1

You must be able to:

- Read values from a real-life graph
- Draw a real-life graph
- Read a conversion graph
- Draw a conversion graph.

Graphs from the Real World

- **Graphs** from the real world include **conversion** graphs.
- You may be asked to convert between these units:
  - £ and $
  - £ and euros
  - pints and litres
  - mph and km/h
  - miles and km
  - gallons and litres

Reading a Conversion Graph

- To convert from one unit to the other, read straight across to the line, then go straight down until you reach the other axis. If you are converting the other way, go up until you reach the line, then read across.

Example
Convert 30 miles into kilometres.

Draw a line **straight up** from 30 miles until it hits the line.

Go **straight across** to the km axis.

30 miles is **equivalent** to 48km.

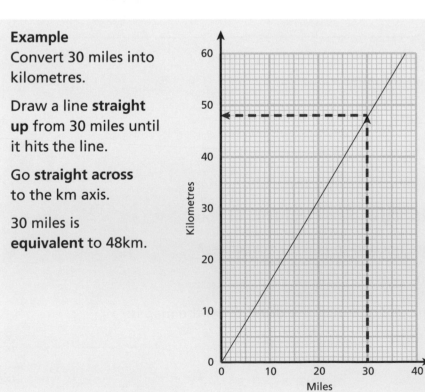

Key Point

To go from kilometres to miles, read straight across to the line, then go straight down until you hit the miles axis.

Drawing a Conversion Graph

- To find the points you need to plot, work out a number of equivalent values.
- Plot these values on a graph and join with a straight line.

Example

Jack's company pays him 80 pence for each mile he travels. Use the information to draw a graph of his pay.

First work out how much Jack will be paid for different journeys.

Distance in miles	0	10	20	30
Amount	0	£8	£16	£24

Now plot the points on a graph and join up with a straight line.

> **Key Point**
>
> Extend your line to the edge of the graph grid.
>
> The graph can now be read to find the pay for different journeys.

Quick Test

Use the conversion graph on page 102 for questions 1 and 2:
1. Find how many kilometres are equivalent to:
   a) 25 miles   b) 10 miles
2. Find how many miles are equivalent to:
   a) 30km   b) 40km
3. Sharon charges $1 for the use of her taxi and 50 cents per mile after that. Work out the cost of a journey that is:
   a) 4 miles   b) 6 miles long
4. Use the information from question 3 to draw a graph of Sharon's charges.

> **Key Words**
>
> graph
> conversion

Real-Life Graphs and Rates 2

You must be able to:

- Read a distance–time graph
- Work out speed, distance and time
- Work out unit prices
- Work out density, mass and volume.

Time Graphs

- Distance–time graphs give information about journeys. Use the horizontal scale for time and the vertical scale for **distance**.
- Distance–time graphs are also used to calculate **speed**.

Example
Amanda cycles to the gym and back every Sunday. The graph below shows Amanda's journey.

Amanda leaves home at 11.20am and cycles 12km in one hour.

Her speed is 12km/h.

Amanda arrives at the gym at 12.20pm and stays for one hour and 20 minutes.

Amanda leaves the gym at 1.40pm and cycles home.

She takes 30 minutes.

Her speed on the return journey is 24km/h.

Travelling at a Constant Speed

- When the speed you are travelling at does not change, it remains **constant**.
- You can work out speed, distance or time using a formula triangle.

Example
A car travels 120 miles at 40 miles per hour. How long does the journey take?

time = distance ÷ speed

time = 120 ÷ 40 = 3 hours

Key Point

$$speed = \frac{distance}{time}$$

Cover up what you are trying to find.

Example

A plane takes $2\frac{1}{2}$ hours to travel 750 miles. What is the speed of the plane?

  speed = distance ÷ time

  speed = 750 ÷ 2.5

    = 300mph

Cover up what you are trying to find.

Unit Pricing

- Unit pricing involves using what 'one' is to work out other amounts. This is often used in conversions of money.

Example

If £1 = $1.75, how much would a pair of jeans cost in $ if they were £60?

  $60 \times 1.75 = \$105$

How much would a TV costing $525 be in £?

  $525 \div 1.75 = £300$

Multiply to get the dollars; divide to get the pounds.

Density

- You can work out **density**, mass and volume using a formula triangle like speed.

  density = mass ÷ volume

Example

Find the density of an object that has a mass of 60g and a volume of 25cm^3.

  density = mass ÷ volume

    = 60 ÷ 25

    = 2.4g/cm^3

Cover up what you are trying to find.

> **Key Point**
>
> Remember to use the correct units.
>
> Volume: **cm^3** and **m^3**
>
> Mass: **g** and **kg**
>
> Density: **g/cm^3** and **kg/m^3**

> **Key Words**
>
> **distance**
> **speed**
> **constant**
> **density**

> **Quick Test**
>
> 1. What is 90 minutes in hours?
> 2. Stuart drives 180km in 2 hours 15 minutes. Work out Stuart's average speed.
> 3. John travelled 30km in $1\frac{1}{2}$ hours. Kamala travelled 42km in 2 hours. Who had the greater average speed?
> 4. If £1 = €1.2, what would £200 be worth in €?
> 5. What is the mass of 250ml of water with density of 1g/cm^3 (1000cm^3 = 1 litre)?

Right-Angled Triangles 1

You must be able to:

- Label right-angled triangles correctly
- Know Pythagoras' Theorem
- Find the length of the longest side
- Find the length of a shorter side.

Pythagoras' Theorem

- Remember the formula for **Pythagoras' Theorem**: $a^2 + b^2 = c^2$

> **Key Point**
>
> The longest side, the hypotenuse, is called c and is opposite the right angle.
>
> The two shorter sides are a and b in any order.

Finding the Longest Side

- To find the longest side **(hypotenuse)**, **add** the **squares**. Then take the **square root** of your answer.

Example

Find the length of y. Give your answer to 1 decimal place.

Label the sides a, b and c first.

Now use the formula:

$$a^2 + b^2 = c^2$$

$$4.1^2 + 13^2 = y^2$$

$$16.81 + 169 = y^2$$

$$185.81 = y^2$$

$$y = \sqrt{185.81}$$

$$= 13.6\text{cm (1 d.p.)}$$

Finding a Shorter Side

- To find a shorter side, **subtract** the squares. Then take the square root of your answer.

Example

Find the length of y. Give your answer to 1 decimal place.

Label the sides a, b and c.

Now use the formula:

$$a^2 + b^2 = c^2$$

$$7^2 + y^2 = 14^2$$

$$49 + y^2 = 196$$

$$y^2 = 196 - 49 = 147$$

$$y = \sqrt{147} = 12.1\text{cm} \ (1 \ \text{d.p.})$$

Example

Find the length of y. Give your answer to 1 decimal place.

Label the sides a, b and c.

Now use the formula:

$$a^2 + b^2 = c^2$$

$$4^2 + y^2 = 15^2$$

$$16 + y^2 = 225$$

$$y^2 = 225 - 16 = 209$$

$$y = \sqrt{209} = 14.5\text{cm} \ (1 \ \text{d.p.})$$

Key Point

You will always $\sqrt{}$ at the end.

Quick Test

1. Work out:
   a) 3.2^2
   b) 15.65^2
2. Work out:
   a) $\sqrt{4900}$
   b) $\sqrt{39.69}$
3. Work out the longest side of a right-angled triangle if the shorter sides are 5cm and 2.2cm.
4. Work out the Shorter side of a right-angled triangle if the longest side is 12cm and the other shorter side is 9cm.

Key Words

Pythagoras' Theorem
hypotenuse
square
square root

Right-Angled Triangles 2

You must be able to:

- Remember the three ratios
- Work out the size of an angle
- Work out the length of a missing side.

Side Ratios

- Label the sides of the triangle in relation to the angle that is marked.

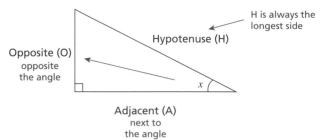

H is always the longest side

Hypotenuse (H)

Opposite (O)
opposite the angle

x

Adjacent (A)
next to the angle

- There are three ratios: **sin**, **cos** and **tan**. Try to find a way of remembering these:

$$\sin x° = \frac{O}{H} \qquad \cos x° = \frac{A}{H} \qquad \tan x° = \frac{O}{A}$$

  – You can use the formula triangles:

  – You can use a rhyme:
    Some **O**ld **H**orses **C**an **A**lways **H**ear **T**heir **O**wners **A**pproach

> **Example**
> Use your calculator to work out the ratios for these angles:
>
> a) $\sin 60° = 0.8660$
> b) $\cos 45° = 0.7071$
> c) $\tan 87° = 19.0811$

Key Point

Ensure your calculator is in **'degree'** mode.

> **Example**
> Use your calculator to work out the angle for these ratios:
>
> a) $\sin x° = 0.5$ $x = \sin^{-1} 0.5 = 30°$
> b) $\cos x° = \frac{3}{5}$ $x = \cos^{-1} (3 ÷ 5) = 53.1°$
> c) $\tan x° = 2.9$ $x = \tan^{-1} 2.9 = 71°$

Finding Angles in Right-Angled Triangles

- You need to know two sides of the triangle to find an angle.

Example
Find the size of angle x. Give your answer to 1 decimal place.

Label the sides O, A and H.

Cross out the side with no information.

As **A** and **H** are left we will use the cos ratio.

Put the information into the triangle.

Cover up what you are trying to find.

$$\cos x° = \frac{3.2}{7}$$

$$x = \cos^{-1}(3.2 \div 7) = 62.8°$$

	Key Point
	The inverse of cos is $\cos^{-1}$.

Finding the Length of a Side

- You need to know the length of one side and an angle to find the length of another side.

Example
Find the length of the side labelled y. Give your answer to 1 decimal place.

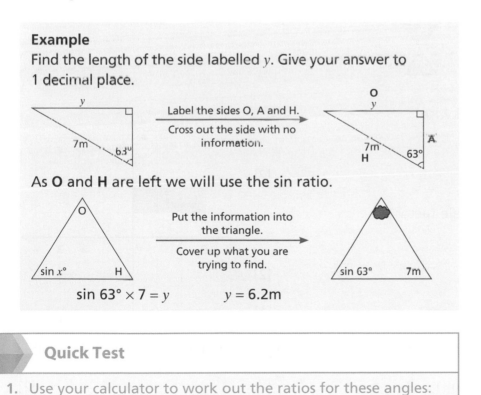

Label the sides O, A and H.

Cross out the side with no information.

As **O** and **H** are left we will use the sin ratio.

Put the information into the triangle.

Cover up what you are trying to find.

$$\sin 63° \times 7 = y \qquad y = 6.2m$$

Quick Test

1. Use your calculator to work out the ratios for these angles:
   a) sin 20° b) cos 30° c) tan 45°
2. Use your calculator to work out the angles for these ratios:
   a) $\sin x° = 0.8337$ b) $\cos x° = \frac{4}{7}$ c) $\tan x° = 32$

	Key Words
	sin
	cos
	tan

Review Questions

Symmetry and Enlargement

PS **1** Reflect shape A across the dotted mirror line.

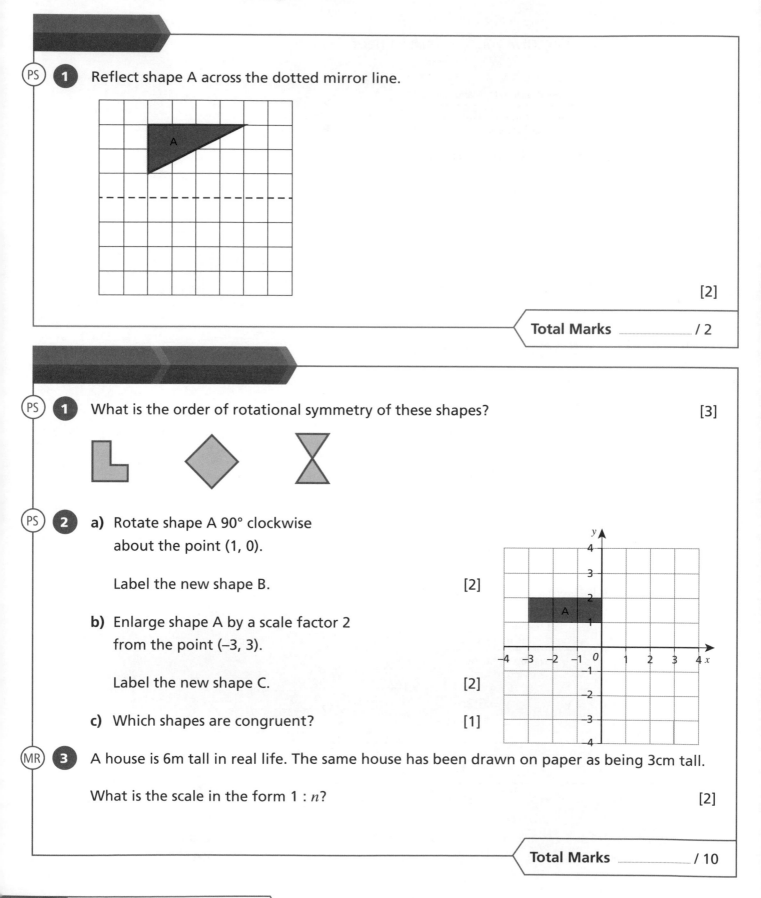

[2]

Total Marks _____ / 2

PS **1** What is the order of rotational symmetry of these shapes? [3]

PS **2** **a)** Rotate shape A 90° clockwise about the point (1, 0).

Label the new shape B. [2]

b) Enlarge shape A by a scale factor 2 from the point (–3, 3).

Label the new shape C. [2]

c) Which shapes are congruent? [1]

MR **3** A house is 6m tall in real life. The same house has been drawn on paper as being 3cm tall.

What is the scale in the form 1 : n? [2]

Total Marks _____ / 10

Ratio and Proportion

PS **1** There are 14 boys and 16 girls in a class.

What is the ratio of girls to boys? Write your answer in its lowest terms. [2]

PS **2** Simplify the following ratios.

a) 10 : 2 [1]

b) 16 : 24 [1]

c) 25 pence : £2 [2]

Total Marks _____ / 6

1 Work out the missing ratio.

a) 8 : 3 – ? : 15 [1]

b) 7 : ? = 63 : 108 [1]

FS **2** A sum of money is shared in the ratio 1 : 4

If the larger share is £120, how much money is there altogether? Show your working. [3]

3 Share 40 pens in the ratio 3 : 5, showing your working. [2]

FS **4** Danesh bought 18 postcards for £2.16

How much would he pay if he bought 27 postcards? Show your working. [2]

PS **5** A telegraph pole 60 feet high casts a shadow 12 feet long. At the same time of day, how long is the shadow cast by a 90-foot pole? [2]

Total Marks _____ / 11

Real-Life Graphs and Rates

1 Concorde could travel 20 miles every minute.

How many miles per hour (mph) is that? [2]

2 Use £1 = €1.19 to work out how much £3.50 is in euros. [2]

Total Marks _____ / 4

1 Each year, there is a tennis competition in Australia and another one in France. The table shows how much money was paid to the winner of the men's competition in each country in one particular year.

Country	Money
Australia	1 000 000 Australian dollars (£1 = 2.70 Australian dollars)
France	780 000 euros (£1 = 1.54 euros)

Which country paid more money? You must show your working. [2]

(PS) **2** The graph shows the flight details of an aeroplane travelling from London to Madrid, via Brussels.

a) What is the aeroplane's average speed from London to Brussels? [2]

b) At what time did the plane arrive in Madrid? [1]

Total Marks _____ / 5

Right-Angled Triangles

(PS) **1** Use your calculator to work out the value of:

    a) 17^2 [1]

    b) 3.5^2 [1]

    c) $\sqrt{529}$ [1]

    d) $\sqrt{40.96}$ [1]

Total Marks _____ / 4

(PS) **1** Use Pythagoras' Theorem to work out:

    a) the length of AC.   **b)** the length of BC.

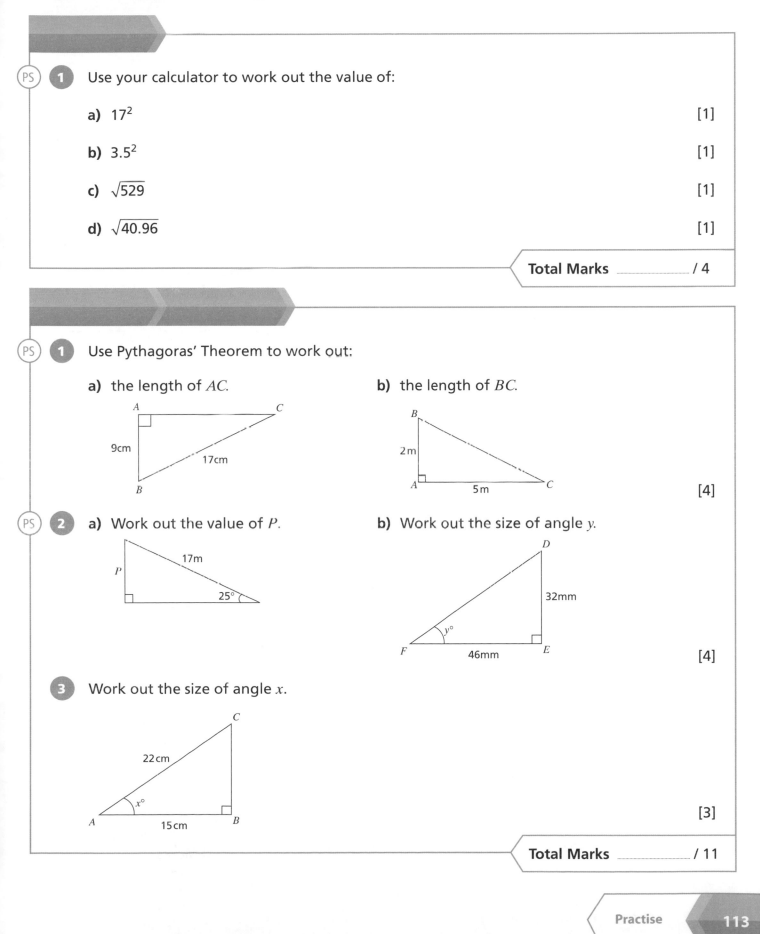

                                                                   [4]

(PS) **2** **a)** Work out the value of P.   **b)** Work out the size of angle y.

                                                                     [4]

3 Work out the size of angle x.

                                                                         [3]

Total Marks _____ / 11

Review Questions

Real-Life Graphs and Rates

(FS) **1** Use £1 = US$1.75 to work out how much:

   a) $200 is in £ [2]

   b) £200 is in US$ [2]

(PS) **2** A coach travels 300 miles at an average speed of 60mph.

   a) For how many hours does the coach travel? [2]

   b) At the same speed, how far will the coach travel in four hours? [2]

Total Marks _____ / 8

(PS) **1** Calculate the density of a piece of metal that has a mass of 2000kg and a volume of $5m^3$. [2]

2 The graphs show information about the different journeys of four people.

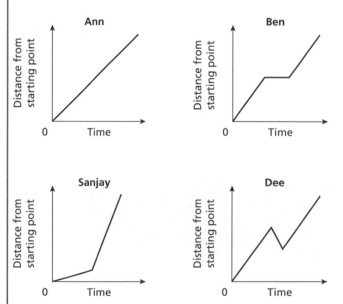

Name	Journey description
	This person walked slowly and then ran at a constant speed.
	This person walked at a constant speed but turned back for a while before continuing.
	This person walked at a constant speed without stopping or turning back.
	This person walked at a constant speed but stopped for a while in the middle.

Write the correct names next to the journey descriptions in the table. [2]

Total Marks _____ / 4

Right-Angled Triangles

(PS) **1** Use your calculator to work out the value of:

  a) 3.3^2 .... [1]

  b) $\sqrt{196}$ .... [1]

(PS) **2** Use Pythagoras' Theorem to work out:

  a) the length of y.

  b) the length of x.

[4]

Total Marks .... / 6

(PS) **1** **a)** Work out the value of angle x.

  b) Work out the length of AB.

[4]

2 An isosceles triangle has a base of length 4cm and a perpendicular height of 8cm.

Giving your answers correct to 1 decimal place, calculate:

  a) the length, y, of one of the equal sides. [3]

  b) the perimeter of the triangle. [1]

Total Marks .... / 8

Mixed Test-Style Questions

No Calculator Allowed 🖩

1 Work out both the surface area and volume of each of these cuboids.

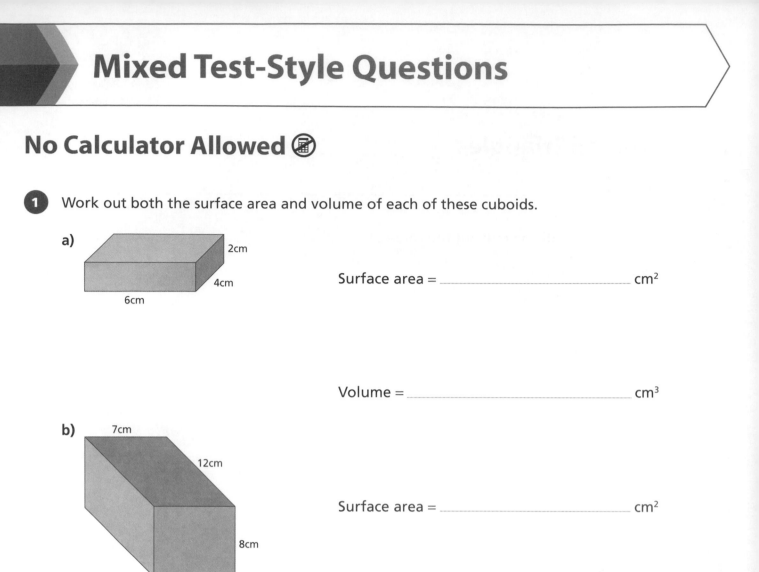

a)

2cm
4cm
6cm

Surface area = _____ cm²

Volume = _____ cm³

b)

7cm
12cm
8cm

Surface area = _____ cm²

Volume = _____ cm³

4 marks

2 Solve the following, giving your answers in the simplest form.

a) $4\frac{1}{2} + 2\frac{1}{3}$

b) $5\frac{2}{3} + 8\frac{1}{4}$

c) $9\frac{1}{6} - 2\frac{3}{8}$

d) $12\frac{1}{2} - 14\frac{5}{6}$

4 marks

3 On the grid below draw the lines for:

a) $y = 6$ b) $x = -4$ c) $y = x$

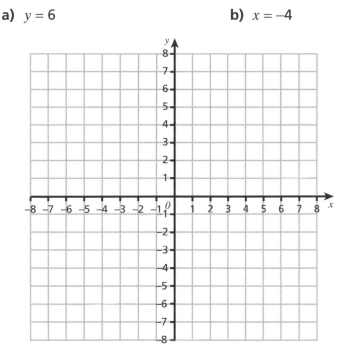

3 marks

4 a) Complete the table of coordinates for the equation $y = 3x - 3$.

x	−2	−1	0	1	2	3
y						

b) Plot the coordinates on the graph below and join them with a line.

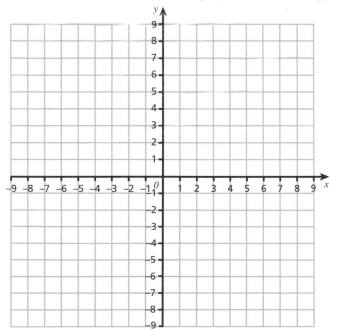

4 marks

TOTAL

15

Mixed Test-Style Questions

5   Calculate angles x and y.

a)

$x =$ _____ °

$y =$ _____ °

b)

$x =$ _____ °

$y =$ _____ °

☐ 4 marks

6   Simplify the following expressions:

a) $3x - 2y + x + 6y$

b) $4g + 5 - g - 4$

☐ 2 marks

7   Expand the following expressions:

a) $4(x - 5)$

b) $4x(x + 4)$

☐ 2 marks

8   Factorise completely the following expressions:

a) $6x - 12$

b) $4x^2 - 8x$

☐ 2 marks

9 The rectangle and trapezium below have the same area.

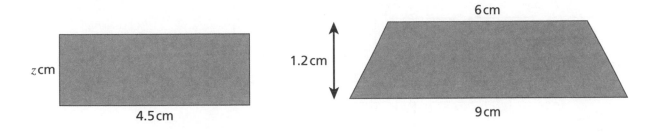

Work out the value of z. Show your working.

10 A chocolate bar costs 60p. A vending machine which sells the chocolate bars is emptied and the following coins are found:

Coins	Frequency
£1	25
50p	59
20p	72
10p	31

How many chocolate bars were sold?

3 marks

TOTAL

16

11 Use the cards below to make two-digit numbers as asked. The first one has been done for you.

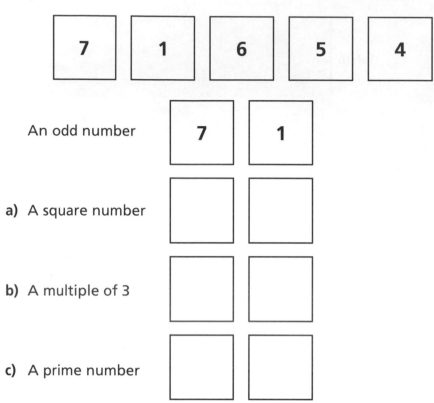

| 7 | 1 | 6 | 5 | 4 |

An odd number | 7 | 1 |

a) A square number

b) A multiple of 3

c) A prime number

2 marks

12 A plant grows by 10% of its height each day. At 8am on Monday the plant was 400mm high.

How tall was it:

a) at 8am on Tuesday?

b) at 8am on Wednesday?

2 marks

13 What are the names of the following shapes?

a)

....

b)

....

c)

....

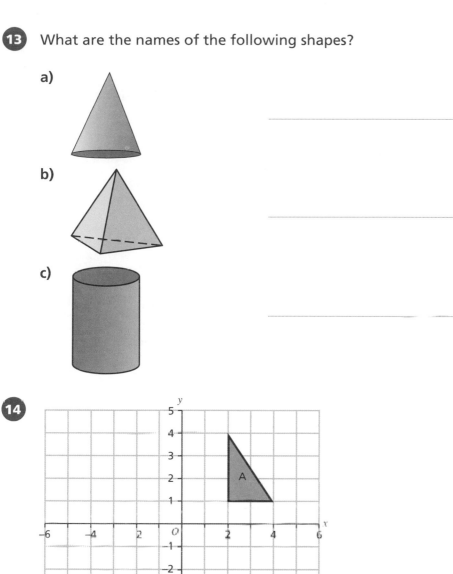

☐ 3 marks

14

a) Reflect shape A in the *y*-axis.

b) Enlarge shape A by a scale factor 2 from the point (3, 4).

c) Rotate shape A 180° about (0, 0).

☐ 3 marks

TOTAL

☐ 10

Calculator Allowed

1 Sam was sitting on the dock of the bay watching boats for an hour. He collected the following information:

Type	Frequency	Probability
tug boat	12	
ferry boat	2	
sail boat	16	
speed boat	10	

a) Complete the table's probability column, giving your answers as fractions.

b) If Sam saw another 75 boats, estimate how many of them would be sail boats.

5 marks

2 Work out the surface area and volume of these cylinders.

a) radius = 4cm

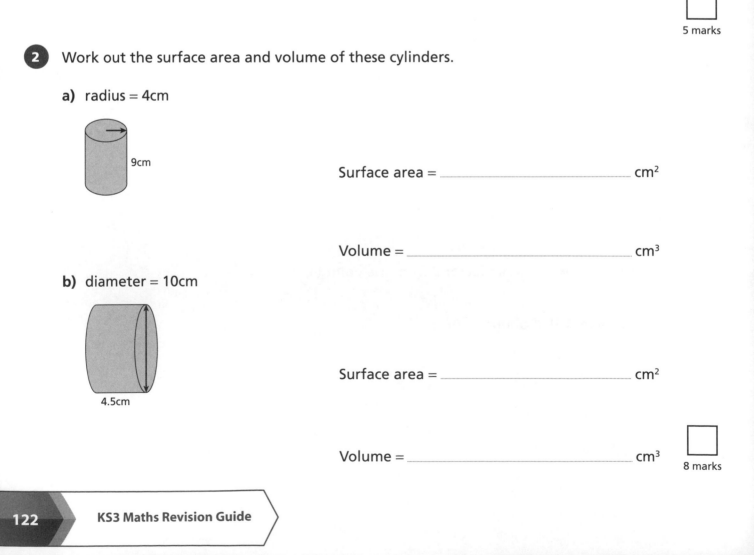

9cm

Surface area = _____ cm²

Volume = _____ cm³

b) diameter = 10cm

4.5cm

Surface area = _____ cm²

Volume = _____ cm³

8 marks

3 The diagram shows a circle inside a square of side length 4cm.

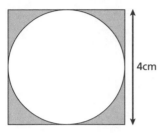

4cm

Find the total area of the shaded regions.

.... cm² ☐

4 Barry is planning to buy a car. He visits two garages which have the following payment options:

Mike's motors	Carol's cars
£500 deposit	£600 deposit
36 monthly payments of £150	12 monthly payments of £50
£150 administration fee	24 monthly payments of £200

Which garage should Barry buy his car from in order to get the cheapest deal? Show working to justify your answer.

☐
3 marks

TOTAL
☐
19

5 An athlete can run 100m in 12 seconds.

Work out the athlete's speed in:

a) m/s

....m/s

b) km/h

....km/h

4 marks

6 A wire 15m long runs from the top of a pole to the ground as shown in the diagram. The wire makes an angle of 45° with the ground.

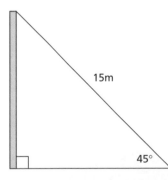

15m

45°

Calculate the height of the pole. Give your answer to a suitable degree of accuracy.

2 marks

7 Below is a map of an island. The scale is 1cm : 4km.

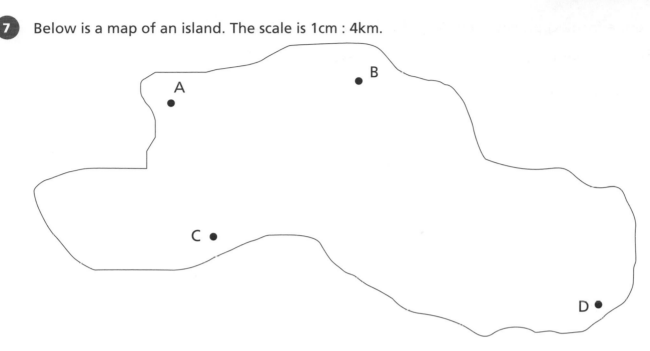

A helicopter flies directly from A to B, B to C, then C to D. What is the total distance flown in km?

.......................km

[ ]

3 marks

8 Change 8% to:

a) a decimal

b) a fraction in its lowest terms

[ ]

2 marks

TOTAL

[ ]

11

9 A recipe for 12 cupcakes needs 80g of butter and 200g of flour.

How much butter and flour are needed to make:

a) 24 cupcakes?

....g of butter

....g of flour

b) 30 cupcakes?

....g of butter

....g of flour

4 marks

10 Work out the missing numbers. You can use the first line to help you.

$16 \times 21 = 336$

a) $16 \times$ .... $= 112$

b) $336 \div 21 =$ ....

2 marks

11 The data below represents the waiting time in minutes of 15 patients in a doctor's surgery:

| 49 | 23 | 34 | 10 | 28 | 28 | 25 | 45 |

| 39 | 35 | 15 | 14 | 48 | 10 | 20 |

a) Draw a stem-and-leaf diagram to show this information.

b) Use your diagram to find the median waiting time.

3 marks

TOTAL

9

Answers

Pages 6–13 Revise Questions

Page 7 Quick Test

1. 3601, 3654, 3750, 3753, 3813
2. 226 635
3. 163
4. 40
5. $-5 < 3$

Page 9 Quick Test

1. 49
2. 7
3. $2^3 \times 5$
4. 252
5. 8

Page 11 Quick Test

1. Output 4 Input 48
2. 21, 25, 29, 33, 37
3. $\div 2$ **OR** $\times \frac{1}{2}$
4. B
5. 39

Page 13 Quick Test

1. 8, 13, 18, 23, 28
2. 5, 8, 13, 20, 29
3. a) $3n + 3$ **OR** $3(n + 1)$ b) 153
4. Position to term rule

Pages 14–15 Review Questions

Page 14

1. 1996 **[1]** because 2000 − 1996 = 4 but 2007 − 2000 = 7   **[1]**
2.
	4	7	6	245	**[2]**
−	2	3	1	**OR** Method mark for valid	
	2	4	5	attempt to subtract	**[1]**

3. 5000, 46 000, 458 000, 46 000   All 4 correct **[2]** Any 3 correct **[1]**
4. $\frac{27}{50}$, 55%, 0.56, 0.6, 0.63
   All 5 correct **[3] OR** 3 out of 5 correct **[2] OR** 55% = 0.55 and $\frac{27}{50}$ = 0.54 **[1]**
5. Ahmed = 2×22 = 44 years old **[2]** or Rebecca = 25 − 3 = 22 years old seen   **[1]**
6. 15 878   **[3]**
   OR 12 000 + 1800 + 210 + 1600 + 240 + 28   **[2]**
   OR Valid attempt at multiplication with one numerical error   **[1]**
7. 52   **[2]**
   OR Valid attempt at division with one numerical error   **[1]**
8.

Both correct **[2]** One correct **[1]**

Page 15

1. 200ml   **[3]**
   OR 1000ml and 800ml seen   **[2]**
   OR 1000ml seen   **[1]**
2. a) 16 **OR** 36   **[1]**
   b) 13 or 17 or 31 or 37 or 53 or 61 or 67 or 71 or 73   **[1]**
   c) 36   **[1]**
   d) 15   **[1]**
3. 9cm **[2] OR** 27 ÷ 3 seen **[1]** (an equilateral triangle has three equal sides)

4. 33   **[2]**
   OR 11 seen   **[1]**
5. 25°   **[3]**
   OR 65 seen   **[2]**
   OR 130 seen   **[1]**
   (angles in a triangle add up to 180°, base angles in an isosceles triangle are equal, right angle is 90°)
6. $T = 200$ **[1]** ($S = T + 100$, $2T + 100 = 500$) $S = 300$ **[1]**

Pages 16–17 Practice Questions

Page 16

1. a) Jan **[1]**   b) Jul **[1]**   c) 12 **[1]** (−1 is below zero)
2. $\sqrt{49}$ = 7cm

1. Jessa is right **[1]** Using BIDMAS multiply is first so
   3 + 20 + 7 = 30 **[1]**
2. a) £4248   **[3]**
   OR 3000 + 500 + 40 + 600 +100 + 8   **[2]**
   OR Valid attempt at multiplication with one numerical error   **[1]**
   b) 354 ÷ 52 = 6.8   **[2]**
   OR Valid attempt at division with one numerical error   **[1]**
   7 coaches needed   **[1]**
   c) 10 spare seats   **[2]**
   OR 364 seen   ($52 \times 7 = 364$)   **[1]**
3. $8^2 = 64$ and $9^2 = 81$ so $\sqrt{79}$ is between 8 and 9   **[2]**
   OR attempt to find any two square numbers each side of 79   **[1]**

Page 17

1. a) 19 **[1]** 23 **[1]**   b) + 4 **[1]**
2. a) 14 **[1]**   b) 9 weeks **[1]**

1. a) $3n + 1$ **[3]**
   OR $3n$ **[2]**
   OR +3 seen as term to term rule   **[1]**
   b) 181   **[1]**
2. 5, 9, 13, 17, 21 = arithmetic. 2, 8, 18, 32, 50 = quadratic.
   8, 17, 32, 53, 80 = quadratric. All three correct **[2] OR** 1 correct **[1]**

Pages 18–25 Revise Questions

Page 19 Quick Test

1. 24cm
2. 27cm²
3. 6cm²
4. Area 19cm²
   Perimeter 24cm

Page 21 Quick Test

1. 16cm²
2. 13cm²
3. Circumference = 37.7cm (1 d.p.) Area = 113.1cm² (1 d.p.)
4. Circumference = 12.6cm (1 d.p.) Area = 12.6cm² (1 d.p.)

Page 23 Quick Test

1. a) Red = 9 Blue = 7 Green = 8 Yellow = 6 Other = 7   b) 37
2. a) 6 and 3   b) mean = 9.2 (1 d.p.) median = 6 range = 37
   c) Median as there is an outlier

Page 25 Quick Test

1. a)

	Football	Rugby	Total
Women	16	9	25
Men	20	10	30
Total	36	19	55

   b) 36   c) 9   d) 55

Pages 26–27 Review Questions

Page 26

1. £550 **[3] OR** £2200 ÷ 4 seen **[2] OR** £2200 seen **[1]**
2. Mercury, Venus, Earth, Mars, Jupiter, Saturn, Uranus, Neptune
   All 8 correct **[2] OR** 6 correct **[1]**
3. 25cm² **[2]**
   OR 5cm seen **[1]**
4. 22 **[2]**
   OR 16 + 6 seen **[1]**

1. 12 48 **[1]** (24 is half of 48) 26 **[1]** (26 is half of 52) 48 **[1]**
2. 15 and 12 **[2] OR** either 15 **OR** 12 seen **[1]**
3. 4 packs of sausages and 3 packs of rolls **[3]**
   OR 24 seen **[2]**
   OR Valid attempt to find LCM of 6 and 8 seen **[1]**

Page 27

1. a) 9 **[1]** 16 **[1]**
   b) **[1]** for each correct answer e.g. $n \div 6$ (or equivalent) or $\sqrt{n}$

1. a) 44 **[1]** (for 20th term $n = 20$) b) 204 **[1]**
   c) $2n - 1$ **[2] OR** $2n$ seen **[1]**
2. 4.03pm **[3] OR** 100 seen **[2] OR** Valid attempt to split 20 and 45 into prime factors seen **[1]** (This question is asking for the LCM)

Pages 28–29 Practice Questions

Page 28

1. Perimeter = 20cm **[1]** Area = 24cm² **[1]**
2. 32cm² **[2] OR** 8 × 4 seen **[1]**

1. $X = 8$cm **[1]** $Y = 6.8$cm **[1]** (area of a rectangle = $l \times w$)
2. a) 192 **[3] OR** 16 × 12 seen **[2] OR** 120 000 and 625 seen **[1]**
   (As each tile is 25cm 16 will fit along one side and 12 along the other)
   b) £300 **[2] OR** 20 seen **[1]** (The nearest multiple of 10 bigger than 192 is 200)
   c) 8 tiles **[1]**

Page 29

1. a) $(6 + 11 + 9 + 12 + 7) \div 5 = 9$ **[1]**
   b) 6, 7, 9, 11, 12 numbers in order, 9 is in the middle **[1]**
   c) Any 5 numbers with a mean of 9 **[1]**
2. 20 **[2] OR** 90 seen **[1]**
3. a) An 8 seen as the last number in the 7 row **[1]**
   b) 59 (remember to look at the key) **[1]**
   c) 37 (a stem-and-leaf diagram puts numbers in order) **[1]**

1. a) Phil 66.4 (1 d.p.) **[1]** Dave 68.9 (1 d.p.) **[1]** (add them up and divide by number of values)
   b) Phil 70 **[1]** Dave 175 **[1]**
   c) Dave as higher average **OR** Phil as more consistent **[2]**

Pages 30–37 Revise Questions

Page 31 Quick Test
1. 235.6
2. 5.6781

3. 100 000
4. 16.2, 16.309, 16.34, 16.705, 16.713

Page 33 Quick Test
1. 49.491
2. 17.211
3. 163
4. 150 000

Page 35 Quick Test
1. $7x + 5y + 6$
2. c^2d^2
3. 14

Page 37 Quick Test
1. $8x - 4$
2. $2x - 14y$
3. $5(x - 5)$
4. $2x(x - 2)$

Pages 38–39 Review Questions

Page 38

1. $\pi \times 49$ **[1]** = 153.9cm² **[1]**
2. Area = 45cm² **[1]**, perimeter = 28cm **[1]**
3. Area = $\frac{1}{2}(6 + 8) \times 4$ **[1]** = 28cm² **[1]**

1. (This is a compound area separated into a rectangle and triangle)
   a) 22m² **[3] OR** 20 and 2 seen **[2] OR** only 20 seen **[1]**
   b) 4 tins **[1]**
   c) £48 **[2] OR** answer to **b** × 12 seen **[1]**
2. Any answer from 5092 to 5095 inclusive (depending on value of pi used) **[2] OR** 157.0796 or 1.57 seen **[1]** (this question is about the circumference of a circle, notice the units are different, 50cm = 0.5m)

Page 39

1.

	Boys	Girls
Right-handed	12	14
Left-handed	7	1

   Completely correct **[3] OR**
   3 boxes correct **[2] OR**
   2 boxes correct **[1]**

2. a)

Vegetable	Frequency
Carrot	4
Peas	8
Potatoes	9
Sweetcorn	3

   All correct **[2] OR** 3 rows correct **[1]**
   b) Potatoes **[1]** (mode is the most common)

1. a) Median **[1]** as data contains an outlier **[1]** (14 808 much bigger than the rest of the data)
   b) 14 808 **[1]**

Pages 40–41 Practice Questions

Page 40

1. $3.7 \times 10 \rightarrow 37$, $3.7 \times 100 \rightarrow 370$, $3.7 \div 10 \rightarrow 0.37$ and $3.7 \div 100 \rightarrow 0.037$. All correct **[2] OR** 1 correct **[1]**
2. 0.5679 greater **[1]** Has 6 hundredths compared to 5 **[1]**

1. a) 57.832 **[1]**
   b) 21.98 **[1]**
   c) 74.154 **[1]**
   d) 216 **[1]**

Page 41

1. $5x = 25$, $x = 9$, $4x = 28$. All correct **[2] OR** 2 correct **[1]**
2. $7y$ **[1]** $12y$ **[1]**

1. Both of them **[1]** $2(x + y)$ expands to $2x + 2y$ **OR** vice versa **[1]**
2. £123 **[3] OR** 48 seen **[2] OR** 120×0.4 seen **[1]**
3. $3x + 7$ **[2] OR** $8x + 2 - 5x + 5$ seen **[1]**
4. $3a(bc + 2)$ **[2] OR** $3(abc + 2a)$ **OR** $a(3bc + 6)$ seen **[1]** (completely means remove all common factors)
5. 144 **[2] OR** 9 seen **[1]** $(ab = a \times b)$
6. $3a - b$, $2a - b$, b. All three correct **[2]** any one correct **[1]**

Pages 42–49 Revise Questions

Page 43 Quick Test
1. Volume = 140cm³
   Surface Area = 166cm²
2. Volume = 440cm³
   Surface Area = 358cm²

Page 45 Quick Test
1. **a)** Volume = 197.9cm³
      Surface Area = 188.5cm²
   b) Volume = 603.2 cm³
      Surface Area = 402.1cm²
2. 174cm³

Page 47 Quick Test
1. 20°
2. Sunday
3. No Helen still uses more (approx. mean 51).

Page 49 Quick Test
1. e.g. money spent on advertising against sales for that item
2. Own goal
3. Any two from: add a time frame, no overlapping tick boxes, cover all outcomes
4. How many sweets do you usually eat in a week?
   plus responses boxes that do not overlap e.g. none /1–2 / 3–4 / more than 4

Pages 50–51 Review Questions

Page 50

1. 0.759 **[2] OR** 253×3 seen **[1]**
2. 17 **[2] OR** $85 \div 5$ seen **[1]**
3. 5.2 and 3.8 circled **[1]**
4. 140 **[2] OR** 7000 and 50 seen **[1]**
5. £854.40 **[2] OR** Valid attempt at multiplication with one numerical error **[1]**

1. 6.765, 6.776, 7.675, 7.756, 7.765: all 5 correct **[2] OR** 3 correct **[1]**
2. **a)** £15.99 **[2] OR** attempt to add up 3 costs with only 1 numerical error **[1]**
   b) £4.01 **[1]**
3. 6.93 **[1]**
4. $-0.5 \leqslant \text{error} < 0.5$ **[1]**

Page 51

1. $11k + 5$ **[1]** $4k + 1$ **[1]**
2. $-2k$ (– sign must be seen) **[1]**
3. cd^2, c^2d, c^2d^2. All three correct **[2] OR** any two correct **[1]**

1. **a)** ab **[1] b)** $2a + 2b$ or $2(a + b)$ **[1] c)** $3a$ and $5a$ **[1]**
2. $4t(2ut - u + 5)$ **[2]**
3. 3600 metres per hour **[2]** $1200 \div \frac{1}{3}$ **OR** 3 seen **[1]**

Pages 52–53 Practice Questions

Page 52

1.

Cuboid	6	12	8	**[1]**
Square-based pyramid	5	8	5	**[1]**
Hexagonal prism	8	18	12	**[1]**

2. Cylinder **[1]**; triangular prism **[1]**; cube **[1]**
3. **a)** 150cm³ **[1]**; 190cm² **[1]**
   b) 81cm³ **[1]**; 117cm² **[1]**

1. $942 \div 5^2 \div \pi$ **[1]** = 12.0cm **[1]**
2. $1385 \div 7^2 \div \pi$ **[1]** = 9.00cm **[1]**

Page 53

1. $\frac{360}{45} = 8$

Category	Frequency	Angle	
Brie	21	$21 \times 8 = 168°$	**[1]**
Cheddar	5	$5 \times 8 = 40°$	**[1]**
Stilton	14	$14 \times 8 = 112°$	**[1]**
Other	5	$5 \times 8 = 40°$	**[1]**
Total	45	360°	

Favourite cheese

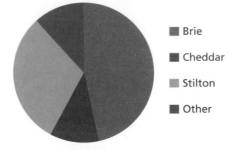

- Brie
- Cheddar
- Stilton
- Other

All angles of pie chart correct to within 2° **[1]**; correct labelling **[1]**

2. Monday has around 55 mins. (Add both Andy's and Helen's phone use together). **[1]**

1. Examples could be: Speed of runner against distance, cost of taxi against number of people in taxi, etc. **[2]**
2. How many more times do you go shopping during the Christmas period than other times of the year?

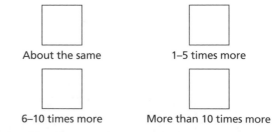

About the same   1–5 times more

6–10 times more   More than 10 times more

Question should have a time frame (e.g. How many more times do you go shopping during the Christmas period than other times of the year?). **[2]**
Response boxes should cover all outcomes and not overlap (e.g. about the same; 1–5 times more; 6–10 times more; more than 10 times more). **[2]**
3. **a)** Scatter graph, frequency polygon, pie chart, bar chart, line graph, histogram, etc. Four examples **[2]**
   b) Choose a graph with a reason, for instance pie chart as it shows the percentage of time spent. **[2]**

Page 55 Quick Test
1. e.g. $\frac{4}{6}$ $\frac{6}{9}$ $\frac{8}{12}$ $\frac{10}{15}$ $\frac{12}{18}$
2. $\frac{64}{77}$
3. $\frac{29}{72}$
4. $\frac{15}{52}$
5. $\frac{81}{100}$

Page 57 Quick Test
1. $\frac{1}{3}$
2. $\frac{49}{36} = 1\frac{13}{36}$
3. $\frac{28}{5} = 5\frac{3}{5}$
4. $\frac{37}{4} = 9\frac{1}{4}$
5. $2\frac{17}{90}$

Page 59 Quick Test
1.

x	−2	−1	0	1	2	3
y	−11	−8	−5	−2	1	4

Page 61 Quick Test
1. a) Gradient = 3
      Intercept = 5
   b) Gradient = 6
      Intercept = −7
   c) Gradient = −3
      Intercept = 2
2.

x	−3	−2	−1	0	1	2	3
y	4	2	2	4	8	14	22

Page 62

1. Square-based pyramid **[1]**
2. a)

[1]

   b)

[1]

c)

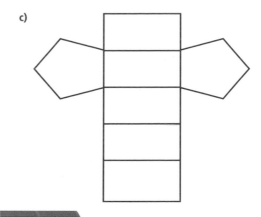

[1]

1. Surface area = 2(14 × 2) + 2(6 × 2) + 2(14 × 6) **[1]** = 248cm² **[1]**
   Volume = 14 × 2 × 6 **[1]** = 168cm³ **[1]**
2. Radius = 2.2 ÷ 2 = 1.1 **[1]**
   $\pi \times 1.1^2 \times 11$ **[1]** = 41.8m³ **[1]**
3. $\sqrt[3]{512}$ **[1]** = 8m = 800cm **[1]**
4. Surface area = 2(15 × 10 + 15 × 20 + 10 × 20) = 1300cm² **[1]**
   No, he has not got enough paper **[1]**

Page 63

1.

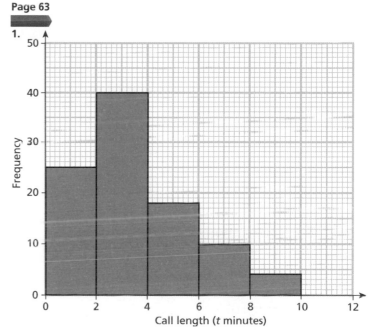

Axes labelled correctly **[1]**; all bars correct **[2]**; three bars correct **[1]**

1.

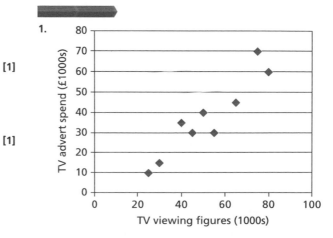

Correct axes labels **[1]**. Points plotted correctly **[1]**. It has a positive correlation **[1]**; the more spent on advertising, the more people are likely to be watching **[1]**

2. a) Example answers: 'a lot' is too vague – the quantity should be specific **[1]**; how much junk food in a period of time could be asked **[1]**

b) Example answers: More answer options could be given, e.g. 0 **[1]**; quantity of fruit could be more specific **[1]**

Page 64

1.
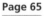

$\frac{3}{5} = \frac{9}{15}$ **[1]** $\frac{2}{3} = \frac{10}{15}$ **[1]**

So $\frac{2}{3}$ is greater than $\frac{3}{5}$ **[1]**

2. a) $\frac{21}{20} = 1\frac{1}{20}$ **[1]**

b) $\frac{41}{40} = 1\frac{1}{40}$ **[1]**

c) $\frac{29}{20} = 1\frac{9}{20}$ **[1]**

d) $\frac{5}{8}$ **[1]**

e) $\frac{3}{10}$ **[1]**

f) $\frac{19}{36}$ **[1]**

3. a) $\frac{2}{24} = \frac{1}{12}$ **[1]**

b) $\frac{40}{54} = \frac{20}{27}$ **[1]**

c) $\frac{3}{20}$ **[1]**

4. a) $\frac{3}{16}$ **[1]**

b) $\frac{9}{48} = \frac{3}{16}$ **[1]**

c) $\frac{21}{12} = \frac{7}{4} = 1\frac{3}{4}$ **[1]**

1. a) $\frac{35}{8} + \frac{11}{5} = \frac{263}{40}$ **[1]** $= 6\frac{23}{40}$ **[1]**

b) $\frac{18}{5} + \frac{21}{9} = \frac{267}{45}$ **[1]** $= 5\frac{14}{15}$ **[1]**

c) $\frac{29}{4} - \frac{30}{11} = \frac{199}{44}$ **[1]** $= 4\frac{23}{44}$ **[1]**

d) $\frac{11}{5} - \frac{10}{7} = \frac{77}{35} - \frac{50}{35}$ **[1]** $= \frac{27}{35}$ **[1]**

Page 65

1. (1, 1) **[1]** and (4, 4) **[1]**

2.
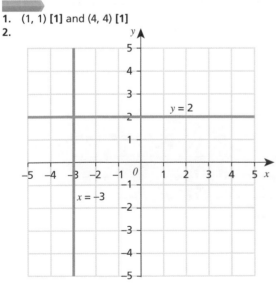

[2]

3.

x	−1	0	1	2	3	4
y	−7	−4	−1	2	5	8

All correct **[2]**; at least four correct **[1]**

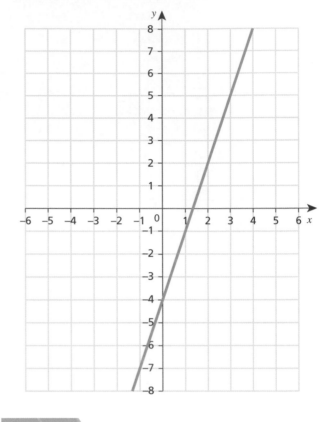

[1]

1.

x	−3	−2	−1	0	1	2	3
y	−2	−4	−4	−2	2	8	16

All correct **[3]**; at least five correct **[2]**; at least three correct **[1]**

Page 67 Quick Test

1. Student's own drawings
2. **a)** 76°
   b) 56°
   c) 65°

Page 69 Quick Test

1. **a)** 55°
   b) 112°
   c) 126°
2. 1440°
3. Equilateral triangle, square or hexagon

Page 71 Quick Test

1. Likely
2.
3. **a)** $\frac{1}{3}$ **b)** $\frac{2}{3}$
4. 0.15

Page 73 Quick Test

1. **a)** 0.4
   b) 0.6
   c) $0.2 \times 60 = 12$

Page 74

1. a) $\frac{3}{4} = \frac{6}{8} = \frac{12}{16} = \frac{60}{80}$ etc. [1]

   b) $\frac{1}{4} = \frac{2}{8} = \frac{4}{16} = \frac{10}{40}$ etc. [1]

   c) $\frac{3}{5} = \frac{6}{10} = \frac{12}{20}$ etc. [1]

1. a) $\frac{4}{10} + \frac{1}{10} = \frac{5}{10} = \frac{1}{2}$ [1]

   b) $\frac{7}{12} + \frac{3}{12} = \frac{10}{12} = \frac{5}{6}$ [1]

   c) $\frac{5}{30} + \frac{6}{30} = \frac{11}{30}$ [1]

   d) $\frac{20}{70} + \frac{21}{70} = \frac{41}{70}$ [1]

   e) $\frac{8}{9} - \frac{3}{9} = \frac{5}{9}$ [1]

   f) $\frac{14}{22} - \frac{11}{22} = \frac{3}{22}$ [1]

   g) $\frac{27}{30} - \frac{20}{30} = \frac{7}{30}$ [1]

2. a) $\frac{4}{45}$ [1]

   b) $\frac{9}{70}$ [1]

   c) $\frac{10}{36} = \frac{5}{18}$ [1]

   d) $\frac{2}{9} \times \frac{4}{1} = \frac{8}{9}$ [1]

   e) $\frac{4}{5} \times \frac{11}{6} = \frac{44}{30} = \frac{22}{15} = 1\frac{7}{15}$ [1]

3. $\frac{2}{5} + \frac{1}{4} = \frac{8}{20} + \frac{5}{20} = \frac{13}{20}$ [1]

   $1 - \frac{13}{20}$ [1] $= \frac{7}{20}$ [1]

4. $\frac{4}{9} + \frac{1}{3} = \frac{7}{9}$ [1] $1 - \frac{7}{9} = \frac{2}{9}$ [1]

   Shared equally $= \frac{1}{9}$ chocolate [1]

5. a) $\frac{77}{9}$ [1]

   b) $\frac{23}{7}$ [1]

   c) $\frac{14}{11}$ [1]

Page 75

1. and 2.

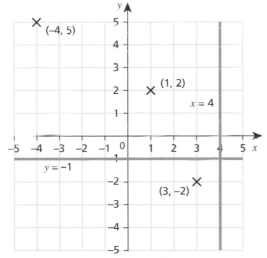

Each point correctly plotted [1]; each line correctly drawn [1]

1.

$\times 2$			-4			
x	-1	0	1	2	3
y	-6	-4	-2	0	2

[2]

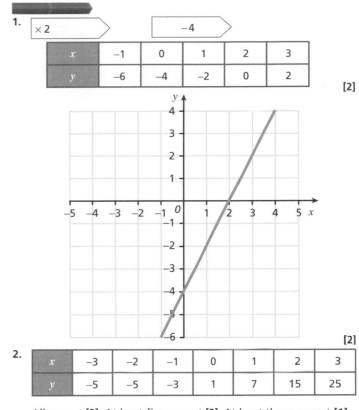

[2]

2.

x	-3	-2	-1	0	1	2	3
y	-5	-5	-3	1	7	15	25

All correct [3]; At least five correct [2]; At least three correct [1]

Page 76

1. a) $180 - 146 = 34°$ [1]

   b) $76°$ (as angles are opposite) [1]

   c) $90 - 56 = 34°$ [1]

2. $180 - 24 = 156°$ [1], $\frac{156}{2} = 78°$ [1]

3. Heptagon [1]

1. a) $x - 134°$ [1] as alternate (Z angle), $y = 180 - 134 = 46°$ [1]

   b) $x = 180 - 53 = 127°$ [1]; $y = 127°$ [1] as it is a corresponding angle (F angle)

2. $180 - 150 = 30°$ [1] (exterior angle)

   $\frac{360}{30} = 12$ sides [1]

Page 77

1. Impossible or 0 [1]

2. a) $\frac{3}{8}$ (number of zeros ÷ total options) [1]

   b) $1 - \frac{1}{8}$ [1] $= \frac{7}{8}$ [1]

1. $1 - 0.65$ [1] $= 0.35$ [1] (Probability of all outcomes take away probability of it landing other way up.)

2. a)

Number	Frequency	Estimated probability
1	5	$\frac{5}{50} = \frac{1}{10}$
2	8	$\frac{8}{50} = \frac{4}{25}$
3	7	$\frac{7}{50}$
4	7	$\frac{7}{50}$
5	8	$\frac{8}{50} = \frac{4}{25}$
6	15	$\frac{15}{50} = \frac{3}{10}$
	Total 50	1

All correct [3]; at least four rows correct [2]; at least two rows correct [1]

b) i) $\frac{3}{10}$ [1]

ii) $\frac{5}{50} + \frac{7}{50} + \frac{8}{50} = \frac{20}{50} = \frac{2}{5}$ [1]

iii) $\frac{15}{50} + \frac{8}{50} = \frac{23}{50}$ [1]

Page 79 Quick Test
1. 0.35 and 35%   2. $\frac{9}{25}$
3. $28   4. $49

Page 81 Quick Test
1. £405
2. £92 000
3. 84%

Page 83 Quick Test
1. ÷ 6   2. 12
3. 5   4. $y = 5$
5. $x = -9$

Page 85 Quick Test
1. $x = 5$ 2. $x = 3$
3. $x = 2$ 4. 11
5. 2kg

Page 86

1. a) $180 - 90 - 46 = 44°$ [1]
   b) $180 - 128 = 52°$ [1]
   c) Because it is an isosceles triangle, $180 - 38 = 142$, $\frac{142}{2} = 71°$ [1]

1. a) $y = 119°$ [1] (as it is corresponding)
   $x = 180 - 61 = 119°$ [1]
   b) $y = 180 - 114 = 66°$ [1]
   $x = 66°$ [1] (as it is an alternate angle)
2. 1260° [1]
3. $180 - 160 = 20°$ [1]
   $\frac{360}{20} = 18$, so it is an 18-sided shape. [1]

Page 87

1. From left: impossible, unlikely, even chance, likely, certain
   All correct [2]; three correct [1]

2. a) $\frac{3}{10}$ [1]
   b) $\frac{4}{10} + \frac{2}{10} = \frac{6}{10} = \frac{3}{5}$ [1]
   c) $1 - \frac{1}{10} = \frac{10}{10} - \frac{1}{10}$ [1] $= \frac{9}{10}$ [1]

1. a)
Sprinkles	Frequency	Probability
Chocolate	19	$\frac{19}{50} = 0.38$
Hundreds and thousands	14	$\frac{14}{50} = 0.28$
Strawberry	7	$\frac{7}{50} = 0.14$
Nuts	10	$\frac{10}{50} = 0.2$

All correct [2]; at least two rows correct [1]

b) $\frac{19}{50} + \frac{10}{50}$ [1] $= \frac{29}{50}$ or 0.58 [1]

2. $1 - 0.47 = 0.53$ [1]

3. a)
Sales destination	Probability of going to destination
London	0.26
Cardiff	0.15
Chester	0.2
Manchester	0.39
[1]

b) Cardiff (it has the lowest probability) [1]

Page 88

1.
Fraction	Decimal	Percentage	
$\frac{7}{10}$	0.7	70	[1]
$\frac{55}{100} = \frac{11}{20}$	0.55	55	[1]
$\frac{32}{100} = \frac{8}{25}$	0.32	32	[1]
$\frac{3}{100}$	0.03	3	[1]

2. a) 50% of £32 = 32 ÷ 2 = £16 [1]
   b) 10% of 80cm = 80 ÷ 10 = 8cm [1]
   c) 10% of 160m = 160 ÷ 10 = 16m
      5% = 16 ÷ 2 = 8m [1]
      15% = 16 + 8 = 24m [1]
   d) 50% = 52
      25% = £26 [1]
3. a) £17 [1]   b) £80 [1]   c) 15m [1]

1. a) 15 ÷ 3 [1]   b) 210 ÷ 7 [1]   c) 6000 ÷ 5 [1]
      5 × 2 [1]   30 × 3 [1]   1200 × 4 [1]
      = £10 [1]   = £90 [1]   = £4800 [1]
2. 10% of £75 = 75 ÷ 10 = £7.50
   20% = £7.50 × 2 = £15 [1]
   Sale price = £75 – £15 [1]
   = £60 [1]
3. Karim gets $\frac{16}{20} = \frac{80}{100}$ [1]
   ↗ ×5
   = 80% [1]
   John gets $\frac{15}{20} = \frac{75}{100}$ [1]
   ↗ ×5
   = 75% [1]

Page 89

1. a) △ = 10 [1]
   b) △ = 5 [1]
   c) $n = 12$ [1]
   d) $y = 7$ [1]
2. a) $3n + 1 = 13$
      (−1) $3n = 12$ [1]
      (÷3) $n = 4$ [1]
   b) $2x - 5 = 3$
      (+5) $2x = 8$ [1]
      (÷2) $x = 4$ [1]
   c) $5y + 1 = 11$
      (−1) $5y = 10$ [1]
      (÷5) $y = 2$ [1]

1. a) $3x + 1 = x + 7$
      (−x) $2x + 1 = 7$
      (−1) $2x = 6$ [1]
      (÷2) $x = 3$ [1]
   b) $2(2x - 3) = x - 3$
             $4x - 6 = x - 3$
      (−x) $3x - 6 = -3$ [1]
      (+6) $3x = 3$
      (÷3) $x = 1$ [1]
   c) $6(x + 1) = 2(x + 13)$
             $6x + 6 = 2x + 26$
      (−2x) $4x + 6 = 26$ [1]
      (−6) $4x = 20$
      (÷4) $x = 5$ [1]

d) $\frac{3x+5}{4} = 5$

(×4) $3x + 5 = 20$ [1]

(−5) $3x = 15$

(÷3) $x = 5$ [1]

2. $3n + 2 = 11$ [1]

(−2) $3n = 9$

(÷3) $n = 3$ [1]

3. $n + 16 = 28$ [1]

(−16) $n = 12$ [1]

Pages 90–97 Revise Questions

Page 91 Quick Test

1. 6

2. a) b) c)

3. Rectangle 3cm × 6cm

Page 93 Quick Test

1. Any three shapes exactly the same size.

2. D 3. 4–5m

4. A and B 5. Any two similar shapes

Page 95 Quick Test

1. a) 6 : 8 b) 8 : 6

2. a) 1 : 3 b) 7 : 1 c) 1 : 4

Page 97 Quick Test

1. 10 : 25 : 5 2. Sara £200, John £160

3. a) 20 b) 28 4. £40

Pages 98–99 Review Questions

Page 98

1. a) $\frac{3}{20} = \frac{15}{100} = 15\%$ [1]

   b) $0.8 = \frac{8}{10}$ [1] $= \frac{4}{5}$ [1]

2. a) $23 \div 2 = £11.50$ [1]

   b) $45 \div 10 = 4.5$cm [1]

   c) $180 \div 4 = 45$m [1]

   d) $10\% = 70 \div 10 = 7$ [1]

   $5\% = 7 \div 2 = £3.50$ [1]

3. a) £14 [1]

   b) £240 [1]

   c) £680 [1]

1. a) $5 \div 5 \times 2$ [1]

   $= £2$ [1]

   b) $£5 - (£2.50 + £2)$ [1]

   $= £0.50$ or 50 pence [1]

2. 10% of £90 = 90 ÷ 10 = £9

   $5\% = £9 \div 2 = £4.50$ [1]

   $15\% = £9 + £4.50 = £13.50$ [1]

   Sale price $= £90 - £13.50$

   $= £76.50$ [1]

3. $150 \div 100 \times 6$

   $6\% = £9$ [1]

   After four years $£9 \times 4 = £36$ [1]

   Total in account = £186 [1]

Page 99

1. a) ○ = 4 [1]

   b) $n = 9$ [1]

   c) $p = 18$ [1]

   d) ○ = 19 [1]

2. a) $4n - 1 = 11$

   (+1) $4n = 12$ [1]

   (÷4) $n = 3$ [1]

b) $5x + 1 = 21$

(−1) $5x = 20$ [1]

(÷5) $x = 4$ [1]

c) $3a + 8 = 5$

(−8) $3a = -3$ [1]

(÷3) $a = -1$ [1]

1. a) $6x - 5 = 4x + 7$

   (−4x) $2x - 5 = 7$

   (+5) $2x = 12$ [1]

   (÷2) $x = 6$ [1]

   b) $5(x + 2) = 2(x - 1)$

   $5x + 10 = 2x - 2$

   (−2x) $3x + 10 = -2$ [1]

   (−10) $3x = -12$

   (÷3) $x = -4$ [1]

   c) $3x - 1 = 4 - 2x$

   (+2x) $5x - 1 = 4$ [1]

   (+1) $5x = 5$

   (÷5) $x = 1$ [1]

2. $\frac{20n + 150}{5} = 50$ [1]

   (×5) $20n + 150 = 250$

   (−150) $20n = 100$

   (÷20) $n = 5$

   The builders worked for 5 hours. [1]

3. $56 - n = 29$ [1]

   $n = 56 - 29 = 27$

   27 chocolate bars were sold. [1]

Pages 100–101 Practice Questions

Page 100

1.

[3]

2. Order 2 [1]

1.

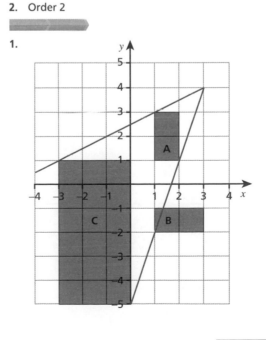

a) See diagram [2]
b) See diagram. [2]
c) A and B [1]
2. 5cm × 4 = 20cm [1]
   7cm × 4 = 28cm [1]

Page 101

1. 3 : 6 **OR** 1 : 2 [1]
2. a) 1 : 3 [1]
   b) 6 : 1 [1]
   c) 20cm : 100cm [1]
      1 : 5 [1]
   d) 80 mins : 90 mins [1]
      8 : 9 [1]

1. 450 ÷ 9 = 50 [1]
   Ann: 4 × 50 = £200 [1]
   Ben: 5 × 50 = £250 [1]
2. 3 parts = £27
   1 part = 27 ÷ 3 = £9 [1]
   2 parts = £9 × 2 = £18 [1]
   Total sum of money = £27 + £18 = £45 [1]
3. Butter: 40 ÷ 6 × 12 = 80g [1]
   Flour: 100 ÷ 6 × 12 = 200g [1]
4. a) 2.5 × 50 000 = 125 000cm [1]
                  = 1250m
                  = 1.25km [1]
   b) 1.4 × 50 000 = 70 000cm [1]
                  = 700m
                  = 0.7km [1]

Pages 102–109 Revise Questions

Page 103 Quick Test
1. a) 40km   b) 16km
2. a) 19 miles   b) 25 miles
3. a) $3   b) $4
4.

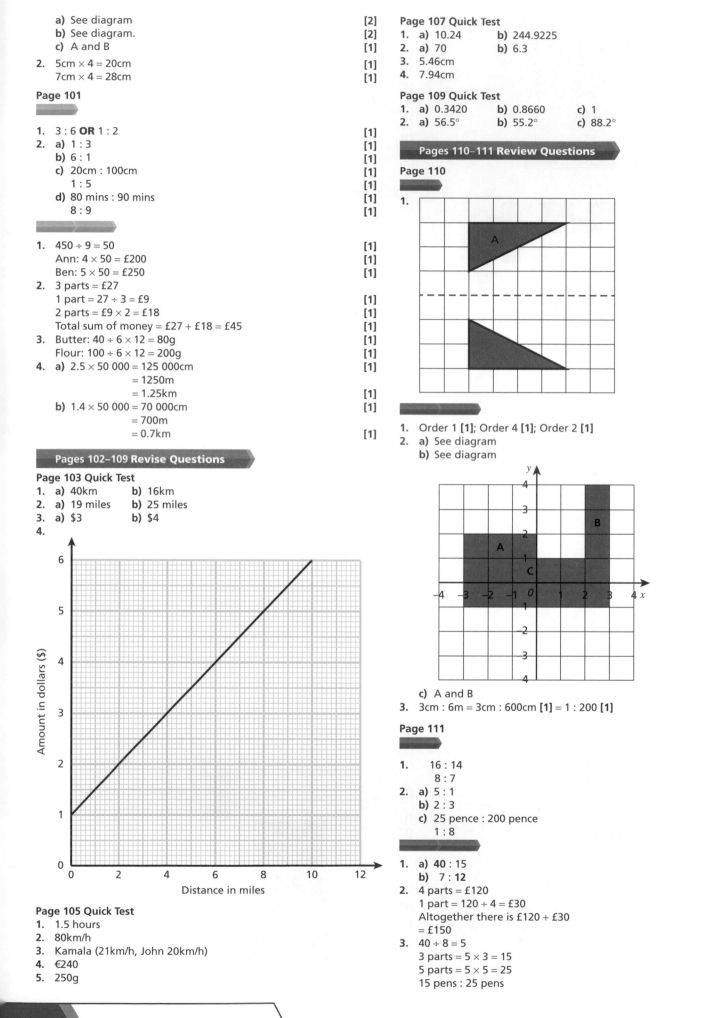

Page 105 Quick Test
1. 1.5 hours
2. 80km/h
3. Kamala (21km/h, John 20km/h)
4. €240
5. 250g

Page 107 Quick Test
1. a) 10.24   b) 244.9225
2. a) 70   b) 6.3
3. 5.46cm
4. 7.94cm

Page 109 Quick Test
1. a) 0.3420   b) 0.8660   c) 1
2. a) 56.5°   b) 55.2°   c) 88.2°

Pages 110–111 Review Questions

Page 110

1.

[2]

1. Order 1 [1]; Order 4 [1]; Order 2 [1]
2. a) See diagram [2]
   b) See diagram [2]

   c) A and B [1]
3. 3cm : 6m = 3cm : 600cm [1] = 1 : 200 [1]

Page 111

1.   16 : 14 [1]
      8 : 7 [1]
2. a) 5 : 1 [1]
   b) 2 : 3 [1]
   c) 25 pence : 200 pence [1]
      1 : 8 [1]

1. a) **40** : 15 [1]
   b) 7 : **12** [1]
2. 4 parts = £120
   1 part = 120 ÷ 4 = £30 [1]
   Altogether there is £120 + £30 [1]
   = £150 [1]
3. 40 ÷ 8 = 5 [1]
   3 parts = 5 × 3 = 15
   5 parts = 5 × 5 = 25
   15 pens : 25 pens [1]

4. 1 postcard = £2.16 ÷ 18 = 12 pence or £0.12   **[1]**
27 postcards cost £0.12 × 27 = £3.24   **[1]**
5. 12 ÷ 60 × 90 **[1]** = 18 foot shadow **[1]**

Pages 112–113 **Practice Questions**

Page 112

1. 20×60 **[1]**
= 1200mph   **[1]**
2. 3.50×1.19 **[1]**
= €4.165   **[1]**

1. France, with correct reasoning, e.g.   **[1]**
Converts $ and € to £
$1\,000\,000 \div 2.7 = £370\,370$ **[1]**
$780\,000 \div 1.54 = £506\,494$
OR
Converts $ to €
$1\,000\,000 \div 2.7 \times 1.54 = 570\,370$
OR
Converts € to $
$780\,000 \div 1.54 \times 2.7 = 1\,367\,532$
2. a) Speed = distance ÷ time
= 350km ÷ 1.1h   **[1]**
= 318km/h   **[1]**
b) Accept 2115 – 2118 (9:15 – 9:18pm)   **[1]**

Page 113

1. a) 289   **[1]**
b) 12.25   **[1]**
c) 23   **[1]**
d) 6.4   **[1]**

1. a) $9^2 + AC^2 = 17^2$
$AC = \sqrt{289 - 81} = \sqrt{208}$ **[1]**
$AC = 14.4$cm   **[1]**
b) $2^2 + 5^2 = BC^2$
$BC = \sqrt{29}$ **[1]**
$BC = 5.39$m   **[1]**
2. a) $\sin 25° = \frac{P}{17}$
$P = \sin 25° \times 17$ **[1]**
$P = 7.18$m   **[1]**
b) $\tan y° = \frac{32}{46}$
$y = \tan^{-1}(32 \div 46)$ **[1]**
$y = 34.8°$ **[1]**
3. $\cos x° = \frac{15}{22}$ **[1]**
$x = \cos^{-1}(15 \div 22)$ **[1]**
$x = 47°$ **[1]**

Pages 114–115 **Review Questions**

Page 114

1. a) 200 ÷ 1.75   **[1]**
= £114.29   **[1]**
b) 200 × 1.75   **[1]**
= US$350   **[1]**
2. a) Time = 300 ÷ 60   **[1]**
Time = 5 hours   **[1]**
b) Distance = 60 × 4   **[1]**
= 240 miles   **[1]**

1. Density = 2000 ÷ 5   **[1]**
Density = 400kg/m³   **[1]**

2.

Name	Journey description
Sanjay	This person walked slowly and then ran at a constant speed.
Dee	This person walked at a constant speed but turned back for a while before continuing.
Ann	This person walked at a constant speed without stopping or turning back.
Ben	This person walked at a constant speed but stopped for a while in the middle.

All correct **[2]**; two correct **[1]**

Page 115

1. a) 10.89   **[1]**
b) 14   **[1]**
2. a) $6^2 + 8^2 = y^2$ **b)** $23^2 + 15^2 = x^2$
$y = \sqrt{100}$ **[1]** $x = \sqrt{754}$ **[1]**
$y = 10$cm **[1]** $x = 27.5$m **[1]**

1. a) $\tan x° = \frac{15}{8}$ **b)** $6^2 + AB^2 = 9^2$
$x = \tan^{-1}(15 \div 8)$ **[1]** $AB = \sqrt{81 - 36} = \sqrt{45}$ **[1]**
$x = 61.9°$ **[1]** $AB = 6.7$cm **[1]**
2. a) $8^2 + 2^2 = y^2$ **[1]** for using '2'
$y = \sqrt{68}$ **[1]**
$y = 8.2$cm **[1]**
b) Perimeter = 8.2 + 8.2 + 4 = 20.4cm (or 20.5 using unrounded values)   **[1]**

Pages 116–127 **Mixed Test-Style Questions**

Pages 116–121 No Calculator Allowed

1. a) Surface area = 2(6 × 4 + 6 × 2 + 4 × 2) = 88cm²   **[1]**
Volume = 6 × 4 × 2 = 48cm³   **[1]**
b) Surface area = 2(12 × 7 + 12 × 8 + 7 × 8) = 472cm²   **[1]**
Volume = 12 × 7 × 8 = 672cm³   **[1]**
2. a) $4\frac{1}{2} + 2\frac{1}{3} = \frac{9}{2} + \frac{7}{3} = 6\frac{5}{6}$ **[1]**
b) $5\frac{2}{3} + 8\frac{1}{4} = \frac{17}{3} + \frac{33}{4} = 13\frac{11}{12}$ **[1]**
c) $9\frac{1}{6} - 2\frac{3}{8} = \frac{55}{6} - \frac{19}{8} = 6\frac{19}{24}$ **[1]**
d) $12\frac{1}{2} - 14\frac{5}{6} = \frac{25}{2} - \frac{89}{6} = -2\frac{1}{3}$ **[1]**

3.

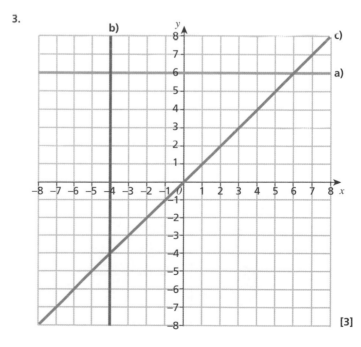

   [3]

4. a)

x	−2	−1	0	1	2	3
y	−9	−6	−3	0	3	6

[2] for all values of y correct; **[1]** for 3 or more correct.

b)

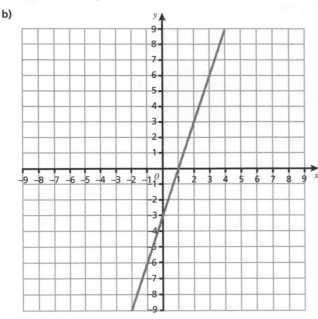

Correctly plotted line **[2]**; Straight line passing through one of the correct coordinates **[1]**

5. a) $x = 115°$ **[1]** (corresponding and opposite), $y = 55°$ **[1]** (alternate)

b) $y = 180 − 85 = 95°$ **[1]**, $x = 180 − 75 = 105°$ **[1]**

6. a) $4x + 4y$ **[1]**

b) $3g + 1$ **[1]**

7. a) $4x − 20$ **[1]**

b) $4x^2 + 16x$ **[1]**

8. a) $6(x − 2)$ **[1]**

b) $4x(x − 2)$ **[1]**

9. $z = 2$ **[3]**; **[2]** if 9 is seen; **[1]** for a correct attempt to find the area of the trapezium with no more than one numerical error.

10. 120 bars **[3]**; **[2]** for $\frac{7200}{60}$ or $\frac{72}{0.6}$ seen; **[1]** for 7200 or 72 seen.

> Remember to work in pounds or pence not a mixture of the two.

11. a) 16 or 64   **b)** 15 or 45 or 51 or 54 or 57 or 75

c) 17 or 41 or 47 or 61 or 67 or 71 **[2]** for all three correct; **[1]** for any two correct.

12. a) $400 + 40 = 440$mm **[1]**

b) $440 + 44 = 484$mm **[1]**

13. a) cone **[1]**   **b)** tetrahedron **[1]**   **c)** cylinder **[1]**

14.

[3]

1. a)

Type	Frequency	Probability	
tug boat	12	$\frac{12}{40} = \frac{3}{10}$	**[1]**
ferry boat	2	$\frac{2}{40} = \frac{1}{20}$	**[1]**
sail boat	16	$\frac{16}{40} = \frac{2}{5}$	**[1]**
speed boat	10	$\frac{10}{40} = \frac{1}{4}$	**[1]**

b) $\frac{2}{5} \times 75 = 30$ **[1]**

2. a) Surface area = 326.7cm² **[2]**; **[1]** for $8 \times \pi \times 9 + 2(4^2 \times \pi)$
Volume = 452.4cm³ **[2]**; **[1]** for $4^2 \times \pi \times 9$

b) Surface area = 298.5cm² **[2]**; **[1]** for $10 \times \pi \times 4.5 + 2(5^2 \times \pi)$
Volume = 353.4cm³ **[2]**; **[1]** for $5^2 \times \pi \times 4.5$

3. 3.43cm² (to 2 d.p.) **[3]**; **[2]** for 16 **and** $\pi(2^2)$ or 12.566 37 seen; **[1]** for 16 **or** $\pi(2^2)$ or 12.566 37 seen.

> Find the area of the square and subtract the area of the circle. Remember the diameter of the circle is the same as the side length of the square, in this case 4cm.

4. Carol's cars and £6000 and £6050 seen **[3]**; **[2]** for £6000 and £6050 seen but no conclusion; **[1]** £5400 seen.

5. a) 8.333m/s **[2]**; **[1]** for $100 \div 12$ seen.

b) 30km/h **[2]**; **[1]** for 0.1km $\div 0.003\,333$h seen.

6. 10.6m **[2]**; **[1]** for $\sin 45° \times 15$

7. 83.6km (±1km) **[3]**; **[2]** for 20km + 22.4km + 41.2km seen; **[1]** for 2 correct measurements in km.

8. a) 0.08 **[1]**

b) $\frac{2}{25}$ **[1]**

9. a) 160g of butter **[1]**, 400g of flour **[1]**

b) 200g of butter **[1]**, 500g of flour **[1]**

10. a) 7 **[1]**   **b)** 16 **[1]**

11. a)

1	0	0	4	5	
2	0	3	5	8	8
3	4	5	9		
4	5	8	9		

Key: $2\,|\,0 = 20$ minutes
Correct diagram with key **[2]**; correct diagram only **[1]**

b) 28 minutes **[1]**

Glossary

a

alternate   angles which are created on a set of parallel lines are the same on a 'Z' angle.

angle   the space (usually measured in degrees) between two intersecting lines or surfaces at or close to the point where they meet.

area   the space inside a 2D shape.

arithmetic sequence   a sequence of numbers with a common difference.

axis   a line that provides scale on a graph; often referred to as x-axis (horizontal) and y-axis (vertical).

b

biased   a statistical event where the outcomes are not equally likely.

bisect   to cut exactly in two.

brackets   symbols used to enclose a sum.

c

centre of enlargement   the position from which the enlargement of a shape will take place.

centre of rotation   the point about which a shape is rotated.

certain   an outcome of an event which must happen, probability equals 1.

chart   a visual display of data.

circle   a round 2D shape.

circumference   the perimeter of a circle.

class interval   the width of the group (difference between the upper and lower limit of the group).

composite or compound   a complex 2D or 3D shape made from several simpler shapes.

congruent   exactly the same.

constant   a value that does not change.

conversion   to change from one unit to another.

coordinates   usually given as (x, y); the x-value is the position horizontally, the y-value the position vertically.

correlation   the relationship between data, the 'pattern'. Can be positive or negative.

corresponding   angles which are created on a set of parallel lines are the same on an 'F' angle.

cos (cosine)   the ratio of the adjacent side to the hypotenuse in a right-angled triangle.

cylinder   a 3D shape with a circular top and base of the same size.

d

data   a collection of answers or values linked to a question or subject.

decimal   a number that contains tenths, hundredths, etc.

decimal places   the number of places after the decimal point.

decimal point   a point used to separate the whole part of a number from the fraction part.

decrease   to make smaller.

degree   the unit of measure of an angle.

denominator   the bottom number of a fraction.

density   the mass of something per unit of volume.

diameter   the distance across a circle, going through the centre.

difference   subtraction.

direct proportion   quantities are in direct proportion if their ratio stays the same as the quantities increase or decrease.

distance   length.

divide   to share.

double   multiply by 2.

e

edge   a line where two faces meet in a 3D shape.

enlargement   a shape made bigger or smaller.

equation   a mathematical statement containing an equals sign.

equivalent   the same as.

estimate   a simplified calculation (not exact), often rounding to 1 significant figure.

even chance   an equally likely chance of an event happening or not happening.

event   a set of possible outcomes from a particular experiment.

expand   remove brackets by multiplying.

experimental probability   the ratio of the number of times an event happens to the total number of trials.

expression   a collection of algebraic terms.

exterior angle   an angle outside a polygon formed between one side and the adjacent side extended.

f

face   a side of a 3D shape.

factor   a number that divides exactly into another number.

factorise   take out the highest common factor and add brackets.

formula   a rule linking two or more variables.

fraction   any part of a number or 'whole'.

frequency   the number of times 'something' occurs.

function machine   a flow diagram which shows the order in which operations should be carried out.

g

geometric sequence   a sequence of numbers in which each term is the product of the previous term multiplied by a constant value.

gradient   the measure of steepness of a line.

graph   a diagram used to display information.

grouped data   data which has been sorted into groups.

h

highest common factor   the highest factor two or more numbers have in common.

hypotenuse   the longest side of a right-angled triangle.

hypothesis   a prediction of an experiment or outcome.

i

impossible   an outcome of an event which cannot happen, probability equals 0.

improper fraction   a fraction where the numerator is larger than the denominator.

increase   to make bigger.

index   the power to which a number is raised; in 2^4 the base is 2 and the index is 4.

integer   whole number.

intercept   the point at which a graph crosses an axis.

interest   an amount added on or taken off.

interior angle   the measure of an angle inside a shape.

interpret   to describe the trends shown in a statistical diagram or statistical measure; the way in which a representation of information is used or surmised.

inverse   the opposite of.

k

key   a statement or code to explain a mathematical diagram.

l

likely   a word used to describe a probability which is between evens and certain on a probability scale.

line of best fit   the straight line (usually on a scatter graph) that represents the closest possible line to each point; shows the trend of the relationship.

linear   in one direction, straight.

lowest common multiple   the lowest multiple two or more numbers have in common.

lowest terms   an answer in its most simplified form.

m

mean   a measure of average; sum of all the values divided by the number of values.

median   a measure of average; the middle value when data is ordered.

mixed number   a number with a whole part and a fraction.

mode   a measure of average; the most common.

mutually exclusive   events that have no outcomes in common.

n

negative   below zero.

net   a 2D representation of a 3D shape, i.e. a 3D shape 'unfolded'.

nth term   see *position to term*.

numerator   the top number of a fraction.

o

ordinary number   a number not written in standard form.

outlier   a statistical value which does not fit with the rest of the data.

p

parallel   lines are said to be parallel when they are at the same angle to one another, and never meet.

parallelogram   a quadrilateral with two pairs of equal and opposite parallel sides.

percentage   out of 100.

perimeter   distance around the outside of a 2D shape.

perpendicular   at 90° to.

pi (π)   the ratio between the diameter of a circle and its circumference, approx. 3.142.

pictogram   a frequency diagram in which a picture or symbol is used to represent a particular frequency.

pie chart   a circular diagram divided into sectors to represent data, where the angle at the centre is proportional to the frequency.

place value   indicates the value of the digit depending on its position in the number.

position to term   a rule which describes how to find a term from its position in a sequence.

positive   greater than zero.

power   see *index*.

prime   a number with exactly two factors, itself and 1.

prism   a 3D shape with uniform cross-section.

probability   the likeliness of an outcome happening in a given event.

probability scale   a scale to measure how likely something is to happen, running from 0 (impossible) to 1 (certain).

product   multiplication.

protractor   a piece of equipment used to measure angles.

Pythagoras' Theorem   in a right-angled triangle, the square on the hypotenuse is equal to the sum of the squares of the other two sides.

q

quadratic   based on square numbers.

quadratic equation   an equation where the highest power of x is x^2.

quadrilateral   a four-sided 2D shape.

quantity   an amount.

r

radius   half the diameter; the measurement from the centre of a circle to the edge.

range   the difference between the biggest and smallest number in a set of data.

ratio   a comparison of two amounts.

raw data   original data as collected.

ray   a line connecting corresponding vertices.

reflection   a mirror image.

regular polygon   a 2D shape that has equal-length sides and angles.

rotation   a turn.

rounding   a number can be rounded (approximated) by writing it to a given number of decimal places or significant figures.

s

sample space   a way in which the outcomes of an event are shown.

scale   the ratio between two or more quantities.

scale factor   the ratio by which a shape/number has been increased or decreased.

scatter graph   paired observations plotted on a 2D graph.

sector   a section of a circle enclosed between an arc and two radii (a pie piece).

sequence   a set of numbers or shapes which follow a given rule or pattern.

share   to divide.

significant figures   the importance of digits in a number relative to their position; in 3456 the two most significant figures are 3 and 4.

similar   two shapes that have the same shape but not the same size.

simplify   make simpler, normally by cancelling a fraction or ratio or by collecting like terms.

simultaneous equations   equations that represent lines that intersect.

sin (sine)   the ratio of the opposite side to the hypotenuse in a right-angled triangle.

solve   work out the value of.

speed   how fast something is moving.

square   a regular four-sided polygon; to multiply by itself.

square number   a number made from multiplying an integer by itself.

square root   the opposite of squaring; a number when multiplied by itself gives the original number.

standard form   a way of writing a large or small number using powers of 10, e.g. 120 000 = 1.2×10^5.

substitute   to replace a letter in an expression with a number.

sum   addition or total.

surface area   the total area of all the faces of a 3D shape.

survey   a set of questions used to collect information or data.

t

tan (tangent)   the ratio of the opposite side to the adjacent side in a right-angled triangle.

term to term   the rule which describes how to move between consecutive terms.

tessellation   a pattern made by repeating 2D shapes with no overlap or gap.

trapezium   a quadrilateral with just one pair of parallel sides.

triangle   a three-sided 2D shape.

u

unlikely   a word used to describe a probability which is between evens and impossible on a probability scale.

units   these define length, speed, time, volume, etc.

v

vertex   the point where the edges meet on a 3D shape.

volume   the capacity, or space, inside a 3D shape.

Index

Notes

KS3 Maths Revision Guide

Collins

KS3
Maths
Foundation Level

Workbook

Samya Abdullah, Rebecca Evans
and Gillian Spragg

Contents

KS3 Maths Workbook

Contents

Number

1 These cards have been placed in the wrong order. They should have been ordered with the smallest number on the left and the largest number on the right.

| 5 | –4 | 0 | –2 | 7 |

Put the cards in the right order.

☐ ☐ ☐ ☐ ☐

[2]

(FS) 2 320 people attended a rock concert. How much money was taken?

Show your working. 🖩

ROCK CONCERT

£6 PER PERSON

£ .... [3]

(MR) 3 Which of these numbers is closer to 500?

497   509

....

Explain how you know.

.... [2]

4 Place one of the symbols between the two numbers to make the statement correct. An example has been done for you.

≤   ≥   <   >

| 5 | < | 8 |
| 9 | | 2 |

[1]

Total Marks .... / 8

(PS) **1** a) Two numbers multiply to make −20. They add to make 1. What are the two numbers?

_____ and _____ [1]

b) Two numbers multiply to make −20, but add to make −1. What are the two numbers?

_____ and _____ [1]

(PS) **2** Tram A leaves the main station every 6 minutes.

Tram B leaves the main station every 14 minutes.

Trams A and B both leave the main station at 9.00am.

When will they next both leave the main station at the same time? Show your working.

_____ _____ [3]

(MR) **3** a) Work out 756 ÷ 21. 🖩

_____ [3]

b) Explain how you would use your answer from part a) to work out 756 ÷ 42.

_____ [1]

(PS) **4** a) Put brackets in the calculation below to make the answer 60.

$5 + 4 \times 7 - 3 = 60$ [1]

b) Put brackets in the calculation below to make the answer −10.

$2 + 3 \times 5 - 7 = -10$ [1]

5 Complete the table to show the inverse operation. The first one has been done for you.

Operation	Inverse operation
Add 2	Subtract 2
Subtract 7	
Multiply by 6	
Divide by 4	

[3]

Total Marks _____ / 14

Sequences

1 The dominoes form a sequence of numbers with a common difference of 4.

| 5 | 9 | | −11 | −7 | | −3 | 1 | | 13 | 17 |

Place the dominoes in the correct order. The first one has been done for you.

| −11 | −7 | | | | | | |

[2]

2 Match the nth term on the left to the first five terms on the right. The first one has been done for you.

3n + 1		2, 7, 12, 17, 22
n + 2		4, 7, 10, 13, 16
5n − 3		3, 4, 5, 6, 7
4n		4, 8, 12, 16, 20

[2]

1 Sarah is organising a gala night and needs to seat 50 people. She is planning to use tables which seat six people, as shown below.

However, Sarah realises when she joins the tables together that they no longer seat six people per table, as shown below.

Sarah works out that she can seat 10 people on two tables and 14 people on three tables.

a) Work out how many tables Sarah needs to seat 50 people.

_____ tables [2]

b) If n represents the number of tables, write down a rule that Sarah could use to work out the number of people that can be seated.

_____ [2]

2 **a)** Find the nth term for this sequence of numbers.

5, 7, 9, 11, 13, _____ [2]

b) Ming thinks this is an arithmetic sequence. Is she correct? Explain your answer.

_____

_____ [2]

3 **a)** Write down the first five terms in the sequence $3n + 2$.

_____ [2]

b) Write down the term-to-term rule.

_____ [1]

4 In the following sequence of numbers, the next term is formed by adding the previous two terms together.

a) Write down the next three terms in the sequence.

1, 1, 2, 3, 5, 8, _____, _____, _____ [1]

b) What is this sequence of numbers commonly known as?

_____ [1]

c) Write down the next three terms in the following sequence of numbers.

1, 4, 9, _____, _____, _____ [1]

d) What is this sequence of numbers commonly known as?

_____ [1]

Total Marks _____ / 15

Perimeter and Area

1 The diagram shows square A.

6cm   A

6cm

I fold square A in half to make rectangle B.

6cm   B

3cm

Then I fold rectangle B in half to make square C.

3cm   C

3cm

a) Complete the table to show the area and perimeter of each shape.

	Perimeter (cm)	Area (cm²)
Square A		
Rectangle B		
Square C		

[3]

b) I fold square A along the diagonal to form a triangle.

6cm

6cm

Work out the area of the triangle. _____ cm² [1]

2 The rectangle has an area of 8cm² and a width of 2cm.

2cm

What is the length of the rectangle? _____ [1]

Total Marks _____ / 5

(MR) **1** Tomiwa and Joe are working out the area of this circle.

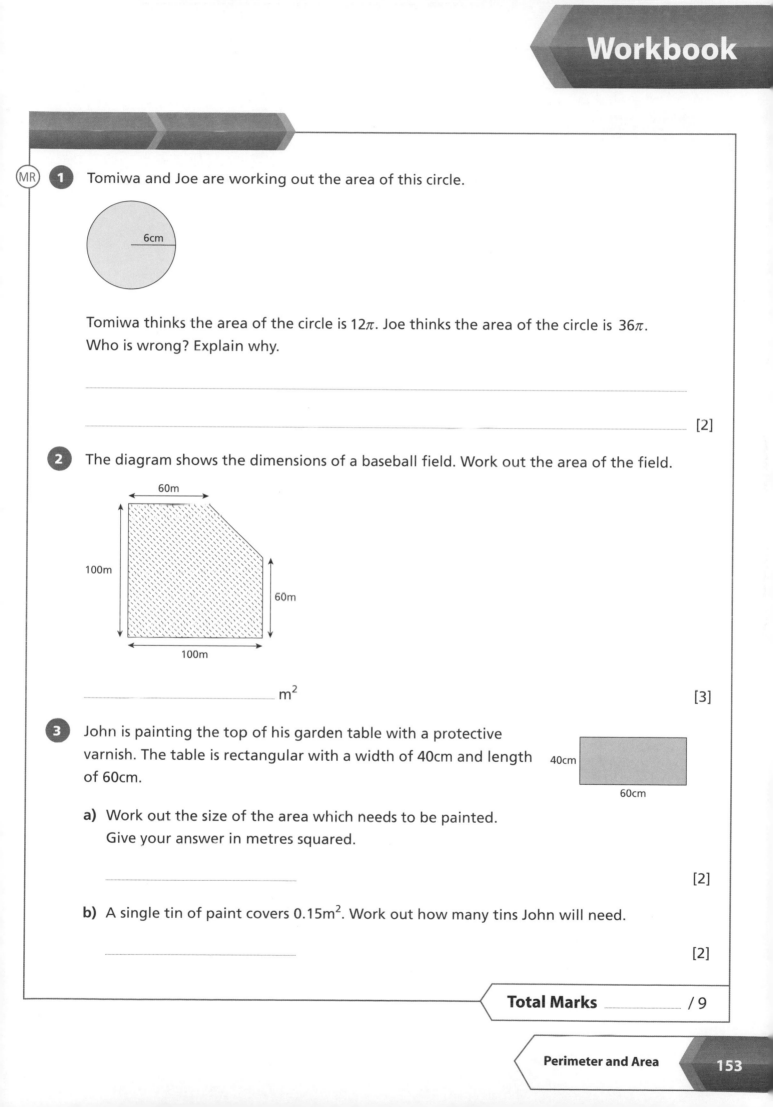

6cm

Tomiwa thinks the area of the circle is 12π. Joe thinks the area of the circle is 36π.
Who is wrong? Explain why.

_____

_____ [2]

2 The diagram shows the dimensions of a baseball field. Work out the area of the field.

60m

100m

60m

100m

_____ m^2 [3]

3 John is painting the top of his garden table with a protective
varnish. The table is rectangular with a width of 40cm and length
of 60cm.

40cm

60cm

a) Work out the size of the area which needs to be painted.
Give your answer in metres squared.

_____ [2]

b) A single tin of paint covers $0.15m^2$. Work out how many tins John will need.

_____ [2]

Total Marks _____ / 9

Statistics and Data

1 | 5 | 4 | 7 | 3 | 2 | 5 | 5 |

a) Write down the mode of the seven numbers shown above.

.... [1]

b) Find the median of the seven numbers shown above.

.... [1]

c) Calculate the mean of the seven numbers above. Give your answer to 1 decimal place.

.... [2]

2 The four pieces of data below have not been entered into the table. Use them to complete the frequency table.

| 25 | 19 | 32 | 45 |

Time	Tally	Frequency
$10 \leqslant t < 20$	ⵏⵏⵏ ⵏⵏⵏ	
$20 \leqslant t < 25$	ⵏⵏⵏ ‖‖‖‖	
$25 \leqslant t < 30$	ⵏⵏⵏ ⵏⵏⵏ ⵏⵏⵏ ‖	
$30 \leqslant t < 50$	ⵏⵏⵏ ‖‖	

[3]

Total Marks .... / 7

1 Look at the data.

6, 6, 6, 8, 8, 8, 9, 9, 10, 11, 12, 15

a) Kirstin thinks the mode is 6, but Dariush says the mode is 8. Who is right?

Kirstin ☐   Dariush ☐   Both of them ☐   [1]

b) What term is used to describe this situation?

.... [1]

(PS) **2** Sarah is investigating the number of hours of sport her friends participate in over a week. Her results are shown below. The median of her results is 6.

2   2   3   6   7   7   9

Sarah asks one more of her friends and her answer is 8.

Does this change the median? Explain your answer.

_____ [1]

3 a) Three numbers have a range of 5, a median of 6 and a mode of 6.

Write down one possibility for the three numbers.

☐ ☐ ☐   [2]

b) Three numbers have a mean of 11 and a range of 2.

Write down the three numbers.

☐ ☐ ☐   [1]

4 a) Calum carries out a survey to find out the most popular choice of music.

Complete the table to show his results.

	Rock	Pop	Classical	Total
Male	18			40
Female			6	
Total		42	13	80

[2]

b) Complete the statements below to help Calum interpret his results. The first one has been done for you.

More women preferred _pop_ than men.

_____ men preferred _____ than women.

_____ was _____ popular than _____.   [2]

Total Marks _____ / 10

Decimals

1 Pair each number on the left with a number on the right to make a **sum** of 2. The first one has been done for you. 🖩

0.02	0.63
0.56	1.98
1.37	1.44
0.51	1.49

[2]

2 Write the correct operation in the box to complete the calculations. The first one has been done for you. 🖩

67	×100	=	6700
5.6		=	5600
876		=	8.76

[2]

Total Marks _____ / 4

1 Pair each number on the left with a number on the right to make a **product** of 2. 🖩

4	0.2
8	0.25
100	0.02
10	0.5

[2]

2 **a)** Write the numbers in order from smallest to largest.

   17.01   17.001   17.91   17.9007

   .... [1]

   b) Work out 1701 × 1791.

   .... [1]

   c) Use your answer to part b) to write down the value of 17.01 × 17.91.

   .... [1]

(MR) **3** 345.6 ÷ 1.5 is the same as 3456 ÷ 15.

Is this statement true? Tick the correct answer.

True [ ]   False [ ]

Give a reason to justify your answer.

....

.... [2]

(FS) **4** Kitchen roll is sold in packs of three and packs of five. A pack of three costs £2.10 and a pack of five costs £3.25.

Which pack gives you better value for money? Show working to justify your answer.

.... [2]

5 Work out 5.26 ÷ 0.02

.... [1]

Total Marks .... / 10

Algebra

1 Two of the three expressions are equivalent.

$3x + 4 + x - 2$ $\qquad$ $5x - 1 - x + 3$ $\qquad$ $2x + 2x + 1 - 3$

a) Circle the odd one out. [1]

b) Justify your answer.

_____

_____ [1]

2 Simplify the expressions.

a) $a \times a \times a$

_____ [1]

b) $b + b + b + b + b$

_____ [1]

c) $c \times c \times d \times d$

_____ [1]

3 Daniel thinks $2(x + 5)$ is equivalent to $2x + 5$.

Explain why Daniel is wrong.

_____

_____ [1]

4 Simplify the expression.

$\frac{2}{3}x + \frac{1}{3}x$

_____ [1]

Total Marks _____ / 7

1 Look at the expressions. Work out the value of each expression when $z = 2$. 🖩

$$4z + 1$$ $$7 + z^3$$ $$3z^2 - 5$$

.... [2]

2 **a)** Find the value of the expression when $x = 2$ and $y = 3$.

$$2xy - 6x^2$$

.... [1]

b) Amy works out the value of the expression in part a) when $x = 4$ and $y = 5$. She also works out the value of the expression below when $x = 4$ and $y = 5$. She gets the same answer for both.

$$2x(y - 3x)$$

Will these two expressions always give the same answer? Give a reason for your answer.

....

.... [2]

3 Multiply out the expressions. Write your answer as simply as possible.

a) $3(y + 2)$

.... [1]

b) $2(v + 1) + 4(v - 2)$

.... [2]

4 Factorise:

a) $10x + 5$

.... [1]

b) $2x^2 + x$

.... [1]

Total Marks .... / 10

3D Shapes: Volume and Surface Area

1 Complete the table.

Shape	Name	Vertices	Edges	Faces
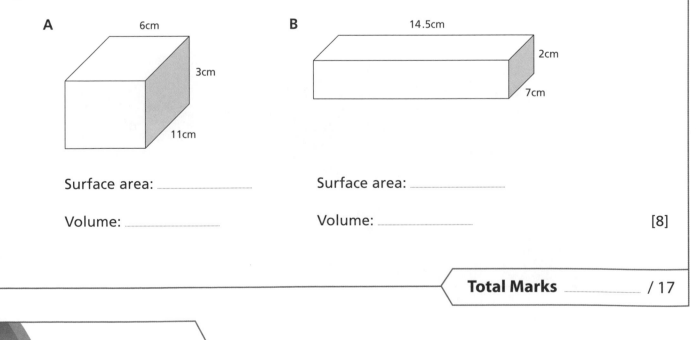				

[6]

2 Draw a net of each of the shapes shown in the table above.

[3]

3 Calculate the surface area and the volume of each of the cuboids.

A 6cm 3cm 11cm

B 14.5cm 2cm 7cm

Surface area: _____

Volume: _____

Surface area: _____

Volume: _____

[8]

1 Calculate the surface area and the volume of each of the cylinders. Show your working.

A

7cm

6cm

B

4cm

25cm

Surface area: ....

Surface area: ....

Volume: ....

Volume: ....   [8]

(PS) **2** What is the height of a cuboid if its volume is 64cm^3, its length is 8cm and its width is 4cm?

....   [2]

(PS) **3** What is the radius of a cylinder that has a volume of 1357.2m^3 and a length of 12m?

12m

?m

....   [2]

(MR) **4** Joseph is buying fruit juice. He has to choose between these two different containers which cost the same.

To get the most juice for his money, which one will he choose? Show your working.

12cm

12cm

10cm

12cm

12cm

Container A   15cm

Container B

....   [4]

Total Marks .... / 16

Interpreting Data

1 60 people were asked what their favourite sweets were. Complete the table and draw a pie chart to show the data.

Category	Frequency	Angle
Polo Mints	8	
Wine Gums	12	
Gummi Bears	17	
Others	23	

[3]

[2]

2 The graph shows Gareth and Rachel's lemonade sales during a week.

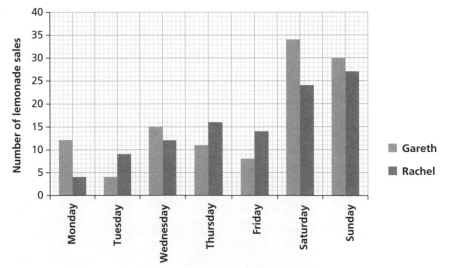

a) On which day was the total lemonade sales the lowest?

.... [1]

b) Who sold the most lemonade throughout the week?

.... [2]

Total Marks .... / 8

1 What two things might you plot against one another to get a positive correlation?

.... [2]

2 Design a question with response box options to determine whether pupils attend the gym more in January than other times of the year.

....

....

....

.... [3]

3 What is similar and what is different about Gareth's and Rachel's lemonade sales in the chart on page 162?

....

....

....

.... [4]

4 Jordan splits his spare time between the hobbies shown in the bar chart.

Does Jordan split his time evenly between musical activities and sporting activities? Justify your answer.

.... [2]

Total Marks .... / 11

Fractions

1 Find the missing values:

a) $\frac{?}{12} = \frac{3}{4}$ .... [1]

b) $\frac{5}{9} = \frac{?}{54}$ .... [1]

c) $\frac{6}{?} = \frac{24}{48}$ .... [1]

2 a) Shade $\frac{5}{6}$ of the shape.

[1]

b) Write $\frac{6}{5}$ as a mixed number.

.... [2]

3 Which fraction is larger, $\frac{3}{4}$ or $\frac{3}{5}$? Use the grids to show your answer.

.... [2]

4 If Sam eats five halves of pizza throughout a weekend, how much pizza does he eat in total? Use the diagrams below to help you.

.... [2]

1 Solve and completely simplify:

a) $\frac{5}{7} + \frac{1}{14} =$ ....

b) $\frac{4}{9} + \frac{3}{12} =$ ....

c) $\frac{2}{5} + \frac{10}{25} =$ ....

d) $\frac{6}{7} - \frac{3}{14} =$ ....

e) $\frac{14}{15} - \frac{3}{5} =$ ....

f) $\frac{9}{10} - \frac{3}{7} =$ .... [6]

2 Solve and completely simplify:

a) $\frac{7}{8} \times \frac{8}{10} =$ ....

b) $\frac{1}{4} \times \frac{5}{6} =$ ....

c) $\frac{5}{8} \times \frac{3}{7} =$ ....

d) $\frac{6}{7} \times \frac{1}{3} =$ ....

e) $\frac{1}{2} \times \frac{5}{8} \times \frac{1}{3} =$ ....

f) $\frac{4}{9} \times \frac{2}{5} \times \frac{6}{15} =$ ....

g) $\frac{7}{10} \times \frac{3}{7} \times \frac{4}{9} =$ ....

h) $\frac{4}{5} \div \frac{1}{2} =$ ....

i) $\frac{3}{7} \div \frac{2}{9} =$ ....

j) $\frac{5}{9} \div \frac{3}{11} =$ .... [10]

(MR) **3** Mathew buys a $4\frac{3}{4}$ kg bag of carrots to make two different soups. One recipe requires $2\frac{1}{3}$ kg of carrots and the other $2\frac{1}{7}$ kg of carrots.

What fraction of carrots is left over?

.... [3]

(MR) **4** Phil has a jug containing $5\frac{3}{10}$ pints of orange juice. He pours $1\frac{1}{2}$ pints of juice into glasses.

How much juice is left in the jug? Give your answer as a mixed number.

.... [3]

Total Marks .... / 22

Coordinates and Graphs

1   Plot these points on the graph.

    a)   (1, 2)   (1, 4)   (3, 4)

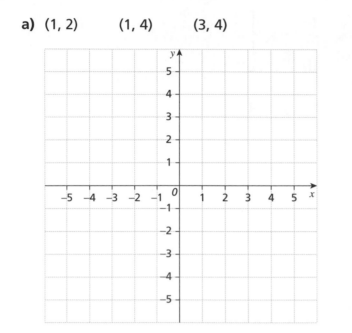

[3]

    b)   Give the pair of coordinates that will complete a square.

      ....   [1]

2   On the graph above plot the lines:

    a) $x = 4$ **b)** $y = -3$ [2]

3   Label the following lines on the graph.

    a) ....

    b) ....

    c) ....

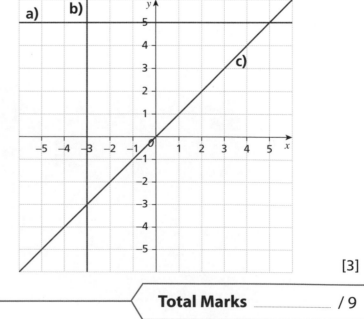

[3]

Total Marks .... / 9

1 **a)** Complete the table for the graph $y = 2x - 1$.

x	−1	0	1	2	3	4
y						

[3]

b) What is the gradient and the y-intercept of this graph?

.... [2]

c) Plot the graph.

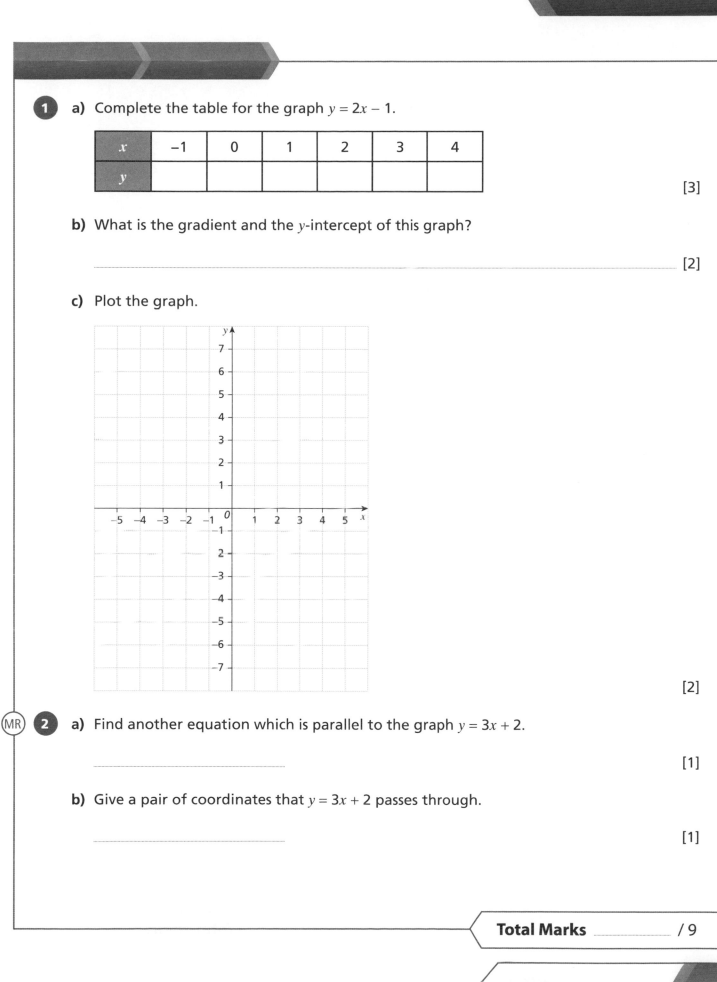

[2]

(MR) **2** **a)** Find another equation which is parallel to the graph $y = 3x + 2$.

.... [1]

b) Give a pair of coordinates that $y = 3x + 2$ passes through.

.... [1]

Total Marks .... / 9

Angles

1 Write down the type of each angle and its size.

a)

.... [2]

b)

.... [2]

c)

.... [2]

d)

.... [2]

2 Pair the angles that make a straight line when put together.

| 23° | 119° | 99° | 157° | 81° | 61° |

.... [3]

3 Using a pair of compasses, bisect the angle below. Show your construction lines.

[3]

Total Marks .... / 14

1 Complete the table.

Number of sides	Name of shape	Sum of interior angles
3		
	Pentagon	
6		
9		

[4]

(MR) **2** Find the missing angles.

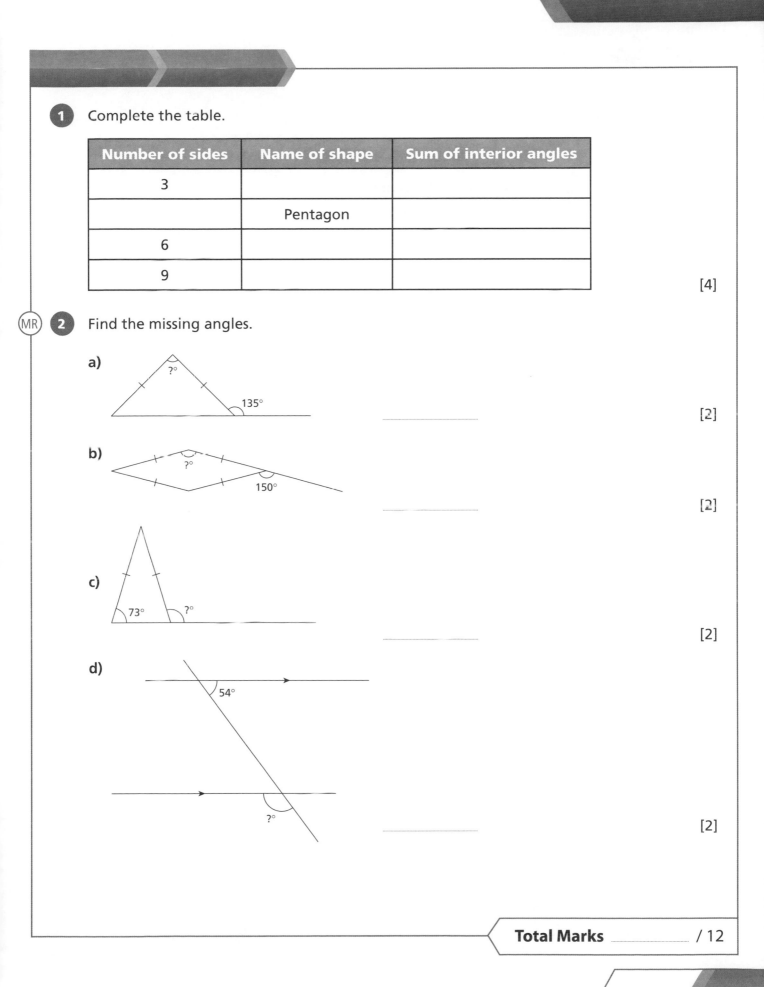

a)

[2]

b)

[2]

c)

[2]

d)

[2]

Total Marks _____ / 12

Probability

1. Put the following events in a suitable place on the scale.

   A You roll a 7 on a normal dice.

   B You toss a coin and get either a head or a tail.

   C Tomorrow you will eat breakfast.

   D You go to school during August.

   E You pick out a red card from a deck of cards.

Impossible	Unlikely	Even chance	Likely	Certain
☐	☐	☐	☐	☐

[5]

2. Jenny has a bag containing 7 purple sweets and 13 green sweets.

   a) What is the probability that she pulls out a purple sweet?

   _____ [1]

   b) What is the probability that she does not pull out a purple sweet?

   _____ [1]

3. The table shows the probability of Hayley's mum giving her a certain lunch.

   a) Complete the table.

Ham sandwich	Cheese sandwich	Chicken wrap	Other
0.3	0.25	0.4	

   [1]

   b) What is the probability that Hayley does not get a ham sandwich?

   _____ [1]

Total Marks _____ / 9

1 Richard has a multipack of crisps containing three bags of cheese and onion, three salt and vinegar, five ready salted and one prawn cocktail. He picks a bag at random.

    a) What is the probability that he takes a salt and vinegar bag?

    .... [1]

    b) If Richard takes a ready salted bag, what is the probability he will take a bag of prawn cocktail next time?

    .... [1]

2 Darren spins his three-sided spinner. The probability that it will land on A is 0.4 and the probability it will land on B is 0.35. What is the probability it will land on C?

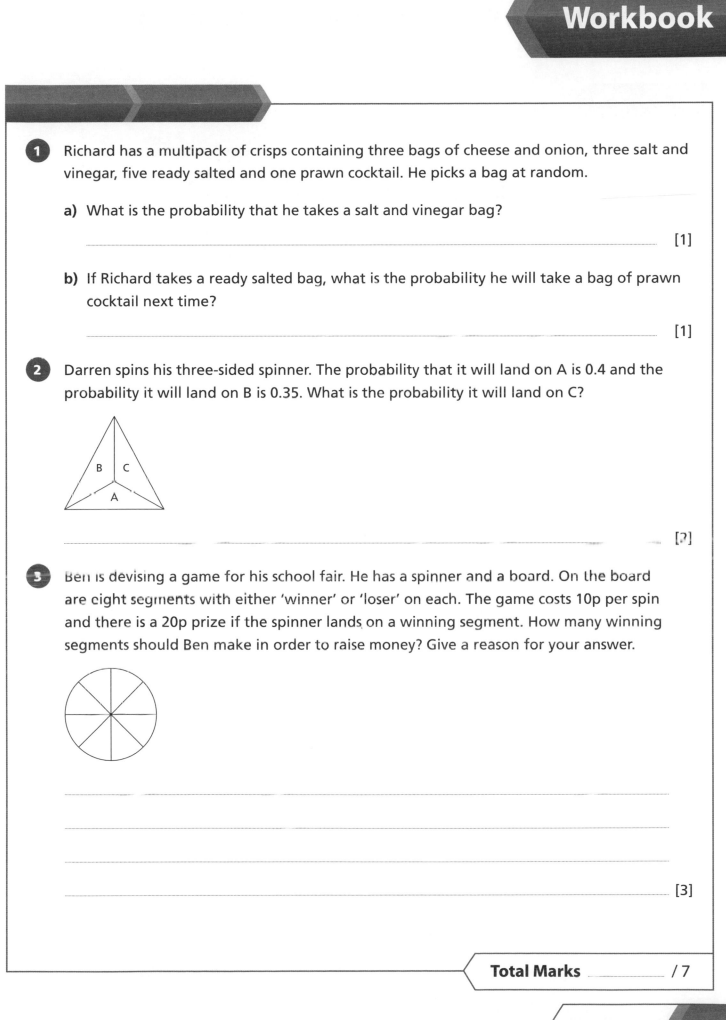

    .... [2]

3 Ben is devising a game for his school fair. He has a spinner and a board. On the board are eight segments with either 'winner' or 'loser' on each. The game costs 10p per spin and there is a 20p prize if the spinner lands on a winning segment. How many winning segments should Ben make in order to raise money? Give a reason for your answer.

    ....

    ....

    ....

    .... [3]

Total Marks .... / 7

Fractions, Decimals and Percentages

(PS) 1 Complete the table of equivalent fractions, decimals and percentages. 🖩

Fraction	Decimal	Percentage
$\frac{3}{10}$		
		45
	0.72	
$\frac{9}{100}$		

[4]

(PS) 2 Change $5\frac{1}{4}$ to a decimal. 🖩

.... [1]

(FS) 3 Work out 25% of £720. 🖩

.... [2]

(PS) 4 Work out $\frac{2}{3}$ of 1500ml. 🖩

.... [2]

(FS) 5 Decrease £430 by 10%. 🖩

.... [2]

(FS) 6 Increase £430 by 20%. 🖩

.... [2]

(PS) 7 A house increases in value from £80 000 to £84 000. 🖩

a) By what fraction does the house increase in value?

.... [2]

b) Write this fraction in its lowest terms.

.... [1]

Total Marks .... / 16

(PS) **1** A certain bacterium grows in number by 10% each day. At midday on Monday there are 50 000 bacteria. How many bacteria will there be: 📱

a) at midday on Tuesday?

_____ [2]

b) at midday on Wednesday?

_____ [2]

(PS) (MR) **2** Anna scored 38 out of 50 in an English test. Ben scored 32 out of 40. Who got the highest percentage?

_____ [2]

(FS) **3** John invests £500 in a building society. He receives 4% simple interest per year. How much will John have at the end of:

a) 1 year?

_____ [2]

b) 3 years?

_____ [2]

(PS) **4** It takes Kamal 45 minutes to get to school and he spends 80% of that time on the bus. How long does Kamal's bus journey take?

_____ [2]

(PS) **5** What percentage of 3kg is 240g?

_____ [2]

(MR) **6** Jane says that $\frac{1}{3}$ is the same as 30%. Jane is wrong.

Explain why Jane is wrong.

_____ [1]

Total Marks _____ / 15

Equations

(PS) **1** Work out what each symbol stands for.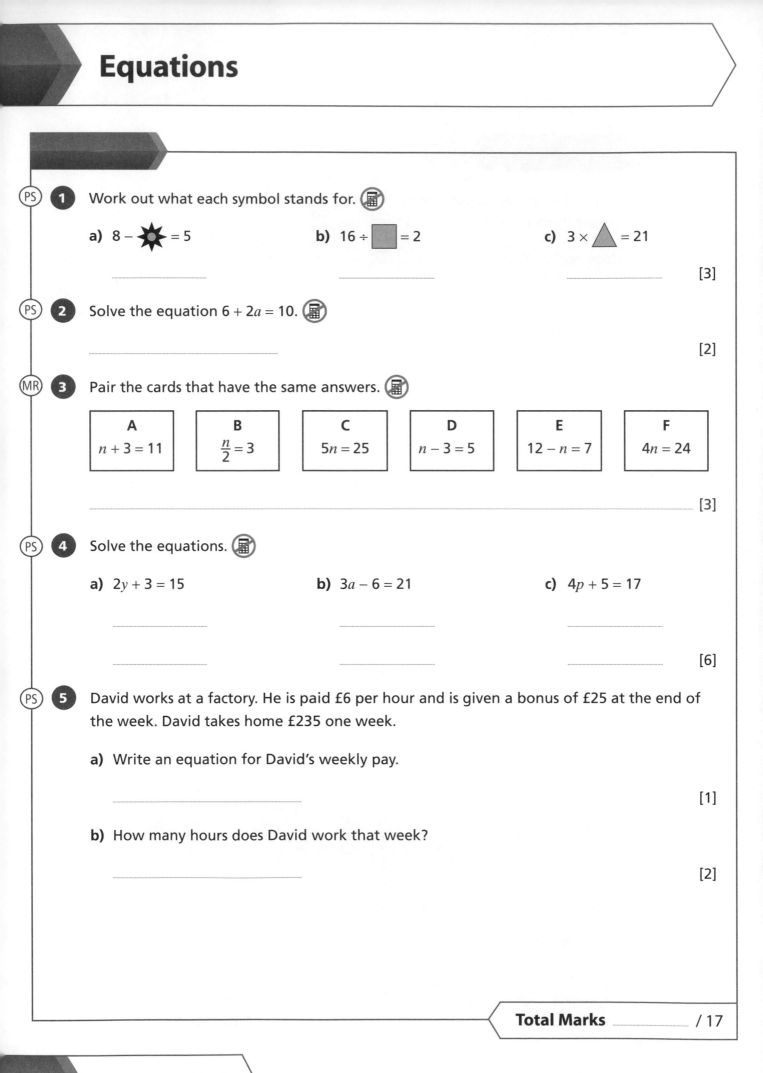

  a) $8 - \text{★} = 5$ **b)** $16 \div \square = 2$ **c)** $3 \times \triangle = 21$

  ....   [3]

(PS) **2** Solve the equation $6 + 2a = 10$.

  ....   [2]

(MR) **3** Pair the cards that have the same answers.

A	B	C	D	E	F
$n + 3 = 11$	$\dfrac{n}{2} = 3$	$5n = 25$	$n - 3 = 5$	$12 - n = 7$	$4n = 24$

  ....   [3]

(PS) **4** Solve the equations.

  a) $2y + 3 = 15$ **b)** $3a - 6 = 21$ **c)** $4p + 5 = 17$

  ....

  ....   [6]

(PS) **5** David works at a factory. He is paid £6 per hour and is given a bonus of £25 at the end of the week. David takes home £235 one week.

  a) Write an equation for David's weekly pay.

  ....   [1]

  b) How many hours does David work that week?

  ....   [2]

Total Marks .... / 17

PS **1** I think of a number, subtract 3 and then multiply the result by 7. My answer is the same as multiplying my number by 4 and then adding 3.

What number am I thinking of?

_____ [3]

PS **2** Solve these equations.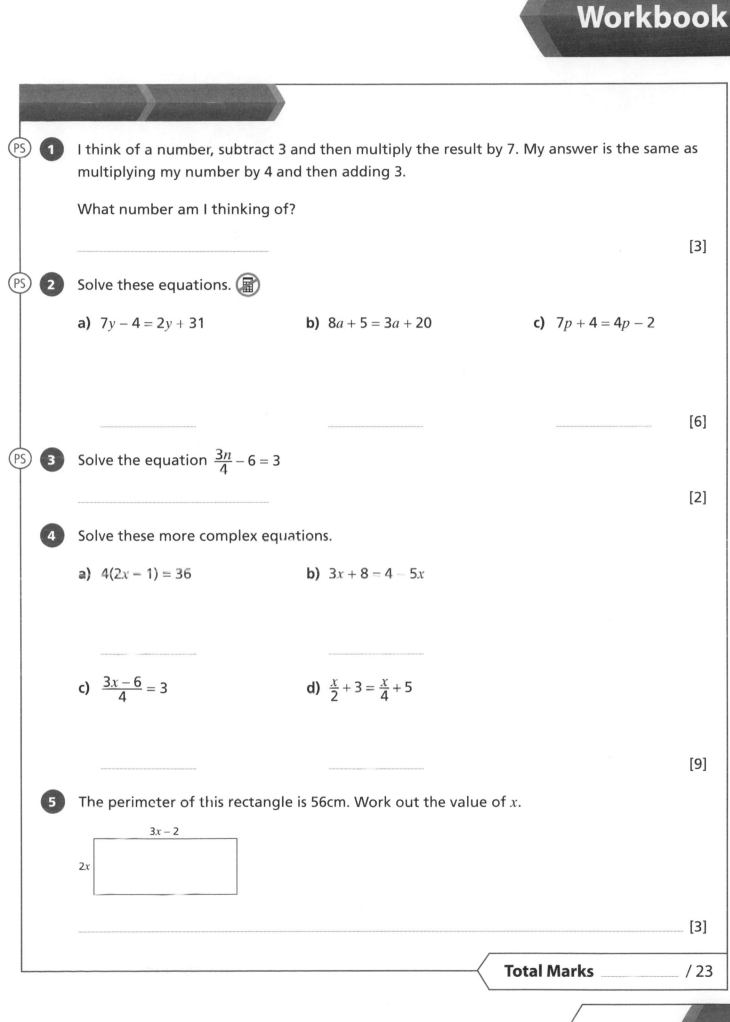

a) $7y - 4 = 2y + 31$ 　　　 b) $8a + 5 = 3a + 20$ 　　　 c) $7p + 4 = 4p - 2$

_____ 　　　 _____ 　　　 _____ [6]

PS **3** Solve the equation $\dfrac{3n}{4} - 6 = 3$

_____ [2]

4 Solve these more complex equations.

a) $4(2x - 1) = 36$ 　　　 b) $3x + 8 = 4 - 5x$

_____ 　　　 _____

c) $\dfrac{3x - 6}{4} = 3$ 　　　 d) $\dfrac{x}{2} + 3 = \dfrac{x}{4} + 5$

_____ 　　　 _____ [9]

5 The perimeter of this rectangle is 56cm. Work out the value of x.

$$3x - 2$$
$$2x$$

_____ [3]

Total Marks _____ / 23

Symmetry and Enlargement

1. These shapes have more than one line of symmetry. Draw the lines of symmetry using a dotted line.

   a)

   b)

   c)

   [3]

(PS) 2. What is the order of rotational symmetry of these shapes?

   a)

   b)

   c)

   ....

   [3]

(PS) 3. Draw in the mirror line that reflects shape A to shape B.

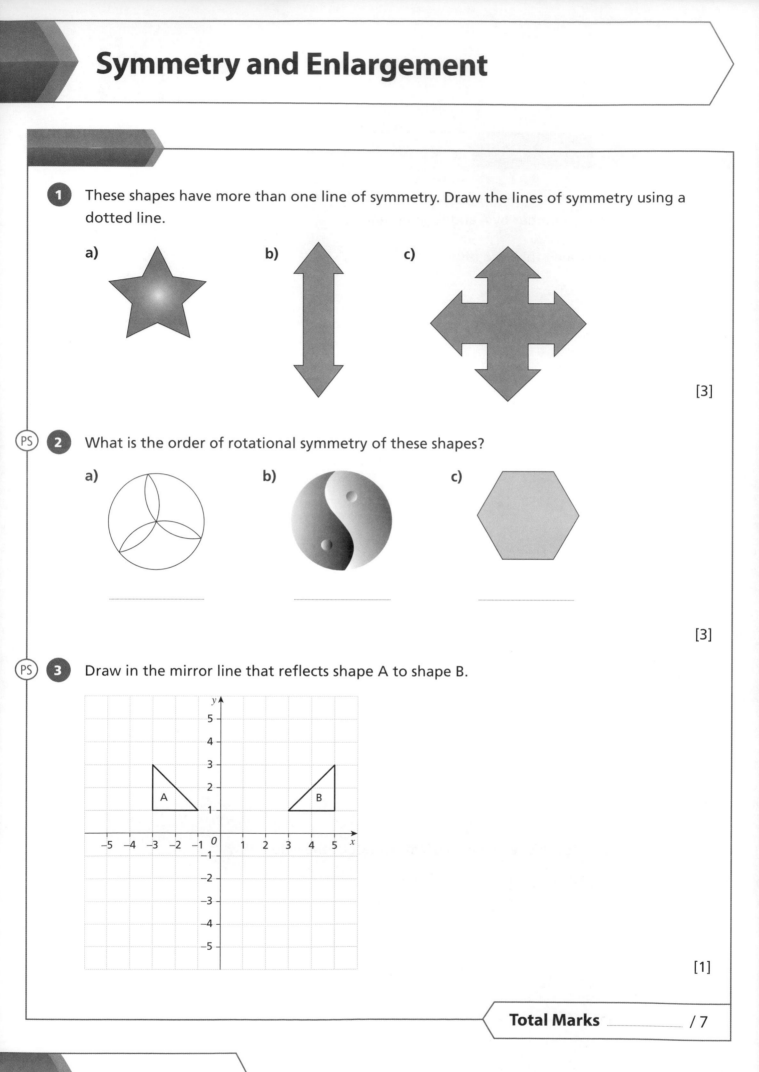

   [1]

Total Marks .... / 7

1

a) Draw the reflection of shape A across the y-axis. Label the image B. [1]

b) Reflect shape B across the x-axis. Label the image C. [1]

c) Describe the rotation needed to get from shape C back to shape A.

_____ [1]

d) Enlarge shape A by a scale factor 2 from the point (−3, 4). [2]

(MR) **2** Here are two models of the same aeroplane.

The wing span of the larger model is 21cm. What is the wing span of the smaller model?

_____ [2]

Total Marks _____ / 7

Ratio and Proportion

PS **1** Write down the ratio of grey squares to white squares. 🖩

.... [1]

PS **2** There are 28 students in a class, 16 of which are boys. 🖩

a) Write down the ratio of boys to girls.

.... [1]

b) Write down the ratio of girls to boys in its lowest terms.

.... [1]

PS **3** Simplify the following ratios. 🖩

a) 28 : 56 .... [1]

b) 18 : 24 .... [1]

c) 40 minutes : 1 hour .... [1]

d) 15mm : 4cm .... [1]

4 Anne weighs 28kg and Ben weighs 40kg. 🖩

a) Write down the ratio of Anne's weight to Ben's weight.

.... [1]

b) Write the ratio in its lowest terms.

.... [1]

Total Marks .... / 9

(PS) **1** Share 700g of sugar in the ratio 3 : 4. 🖩

_____ [2]

(FS) **2** An amount of money is shared in the ratio 2 : 5. 🖩

a) If the smaller share is £16, how much is the larger share?

_____ [2]

b) What is the total amount to be shared?

_____ [2]

(PS) **3** My recipe makes 12 cupcakes. 🖩

a) How much flour will I need to make 24 cupcakes?

_____ [1]

b) How many eggs will I need to make 24 cupcakes?

_____ [1]

c) How much sugar will I need to make 30 cupcakes?

_____ [1]

Recipe for 12 cupcakes
100g butter
100g caster sugar
100g self-raising flour
$\frac{1}{2}$ teaspoon baking powder
$\frac{1}{2}$ teaspoon vanilla essence
20ml milk
2 eggs

(PS) **4** If 8 pencils cost £1.60, how much will 3 pencils cost?

_____ [2]

(PS) **5** A sheet of gift wrap is 1.2m long. It is cut into two pieces, the shorter piece being 45cm in length. Find the ratio of the lengths of the two pieces, giving your answer in its lowest terms. 🖩

_____ [3]

Total Marks _____ / 14

Real-Life Graphs and Rates

1 You can use this graph to convert miles into kilometres.

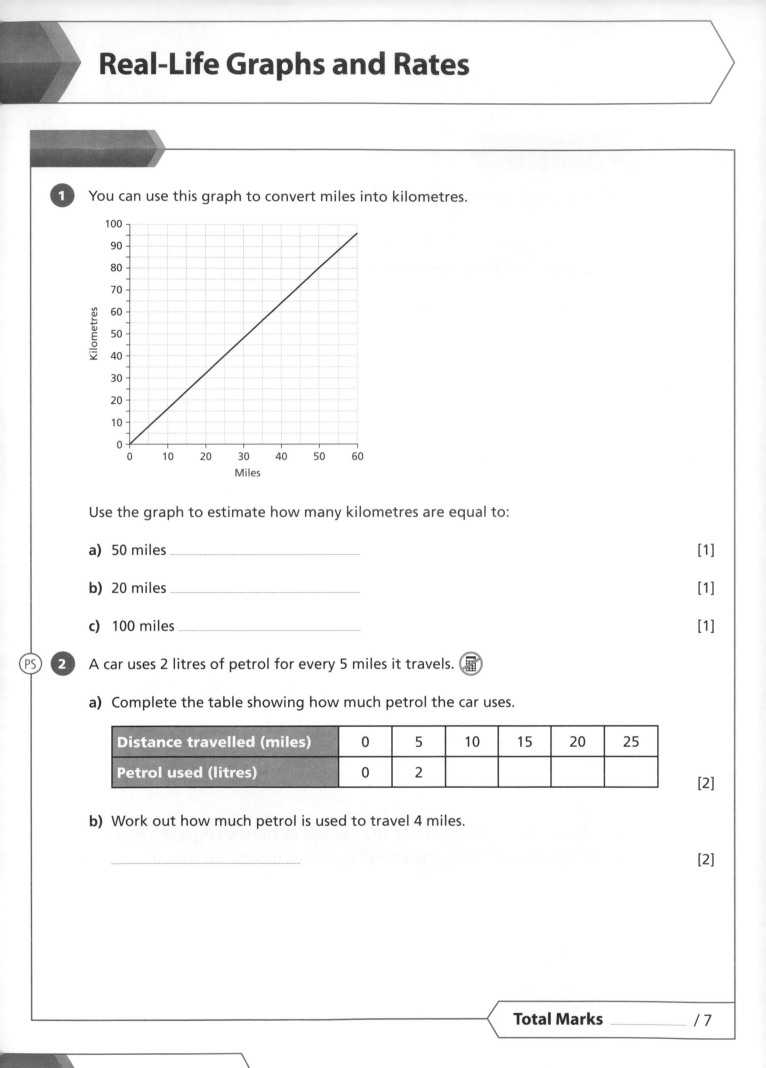

Use the graph to estimate how many kilometres are equal to:

a) 50 miles _____ [1]

b) 20 miles _____ [1]

c) 100 miles _____ [1]

(PS) **2** A car uses 2 litres of petrol for every 5 miles it travels.

a) Complete the table showing how much petrol the car uses.

Distance travelled (miles)	0	5	10	15	20	25
Petrol used (litres)	0	2				

[2]

b) Work out how much petrol is used to travel 4 miles.

_____ [2]

1. This graph shows Jason's car journey to visit his grandmother. He stops for a break on the way there.

a) How far away does his grandmother live?

_____ [1]

b) How long does it take Jason to drive to his grandmother's house?

_____ [1]

c) How long does Jason stay at his grandmother's house?

_____ [1]

d) Work out Jason's average speed on his return journey.

_____ [2]

2. A bus travels at a speed of 30 miles per hour for 4 hours. How far does it travel? Give your answer in miles.

_____ [2]

(FS) 3. Use £1 = €1.22 to work out how much £200 is in euros.

_____ [2]

(FS) 4. Use £1 = €1.22 to work out how much €200 is in pounds.

_____ [2]

5. The density of oil in a tank is 700kg/m^3. The mass of the oil is 15 tonnes.

Calculate the volume of the oil in the tank. (1 tonne = 1000kg)

_____ [2]

Total Marks _____ / 13

Right-Angled Triangles

1 Use Pythagoras' Theorem to work out the values of p and q.

a)

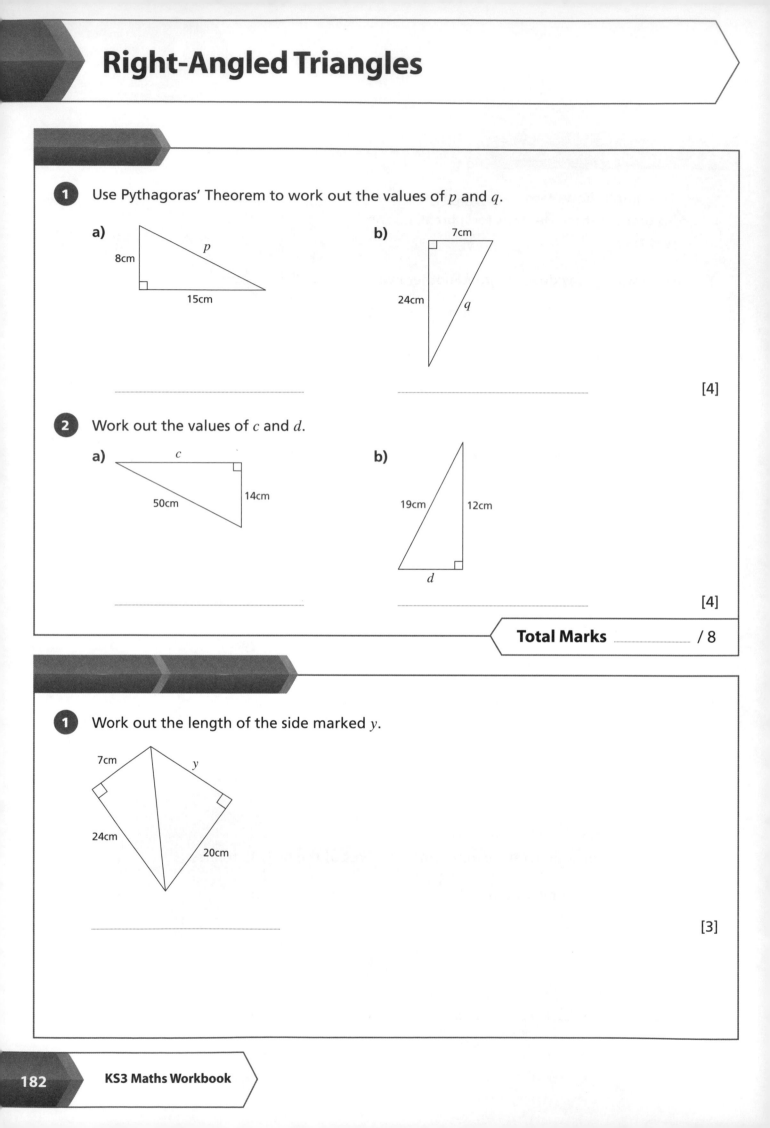

8cm

p

15cm

b)

7cm

24cm

q

.... [4]

2 Work out the values of c and d.

a)

c

50cm

14cm

b)

19cm   12cm

d

.... [4]

Total Marks .... / 8

1 Work out the length of the side marked y.

7cm y

24cm

20cm

.... [3]

2 Find the value of the following to 2 decimal places.

a) $5\sin 10°$ _____ (2 d.p.) [1]

b) $20\cos 34°$ _____ (2 d.p.) [1]

c) $\dfrac{15}{\tan 35°}$ _____ (2 d.p.) [1]

3 Work out the length of the side marked x. Give your answers to 1 decimal place.

a)

67°
x
15cm

_____ (1 d.p.)

b)

x
22cm
12°

_____ (1 d.p.) [4]

4 Work out the size of the angle marked x. Give your answers to 1 decimal place.

a)

57cm
x
39cm

_____ (1 d.p.)

b)

x
8cm
12cm

_____ (1 d.p.) [4]

5 A ladder which is 5m long is placed against a wall.
It makes an angle of 44° with the wall.

Wall
44°
5m
Ladder

a) How high up the wall does the ladder reach?

_____ [2]

b) How far away from the wall is the base of the ladder?

_____ [2]

Total Marks _____ / 18

Mixed Test-Style Questions

No Calculator Allowed 🖩

1 Marissa logs how many emails she receives each day for 20 days. She records the data in a stem-and-leaf diagram.

0	4	4	7			
1	3	3	7	7	9	
2	5	5	7	8	8	8
3	1	1	5	7		
4	0	6				

Where **2 | 4 = 24,**

a) i) what is the mode number of Marissa's emails?

....

ii) what is the median?

....

b) What other values can you calculate from the diagram?

....

[ ] 4 marks

2 Solve and fully simplify the following.

a) $\frac{3}{7} + \frac{2}{5} =$ ....

b) $\frac{3}{8} + \frac{1}{9} =$ ....

c) $\frac{4}{7} - \frac{1}{10} =$ ....

[ ] 3 marks

3 Share £840 in the ratio 3 : 4 : 5.

....

[ ] 3 marks

4 Look at the calculation.

$5^2 + 10^2 = 5^x$

Find the value of x.

....

[ ] 2 marks

5 **a)** Plot these lines on the graph.

    i) $y = 4$ **ii)** $x = 5$ **iii)** $y = x$

  b) Plot these coordinates on the graph.

    (2 , 5) (1 , 1)

  c) Create an isosceles triangle by finding a third coordinate in the positive quadrant of the grid.

    ....

6 marks

---

6 Robert buys a cereal selection pack. The contents are shown in the table.

Honey Crunch	Lucks	Frostables	Coco Bangs	Special J
3	2	4	4	1

  a) What is the probability that Robert selects (at random) a pack of Coco Bangs?

    ....

  b) Which cereal is Robert least likely to select?

    ....

  c) What is the probability that Robert will not select a pack of Lucks?

    ....

4 marks

7 John invests £200 in a building society. He receives 4% simple interest per year. How much will John have at the end of 3 years?

    ....

2 marks

TOTAL

24

Mixed Test-Style Questions

8 Write the shaded area of these shapes as a fraction and a percentage.

a)

Fraction = ...

Percentage = ...

b)

Fraction = ...

Percentage = ...

9 This is an isosceles triangle. The perimeter of the triangle is $5a + 4b$.

a) Find an expression in terms of a and b for the length of side AB.

....

b) Given that the value of a is 6 and the perimeter is 62, find the value of b.

....

10 Translate this shape by vector $\begin{pmatrix} -3 \\ 2 \end{pmatrix}$.

11

	£10	£40	£65
10%	£1	£4	£6.50
5%	50p	£2	£3.25

Use the table to complete the following.

a) 15% of £65 = _____

b) £1.50 = 15% of _____

c) £14 = _____ of £40

d) £1 = 5% of _____

4 marks

12 In the diagram below three squares in a straight line must add up to 150.

Fill in the missing numbers.

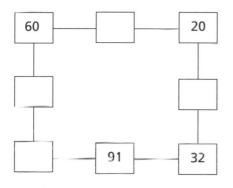

2 marks

13 Which is greater, 2^3 or 4^2? Show working to justify your answer.

_____

1 mark

14 A man cycles 10 miles in half an hour.

What is his speed? Give your answer in miles per hour.

_____

2 marks

TOTAL

19

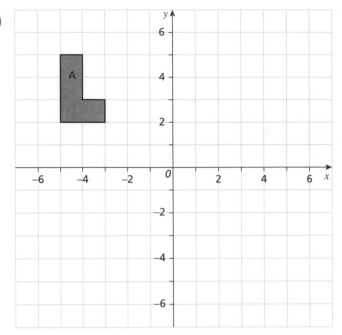

a) Reflect shape A across the x-axis.

b) Enlarge shape A by a scale factor 3 from the point $(-6, 5)$.

c) Rotate shape A 180° from $(0, 0)$.

5 marks

16 The length of a rectangle is 6cm more than its width. The perimeter of the rectangle is 40cm. Work out the width (x).

$x + 6$

x

....

3 marks

17 Paul has a collection of CDs of which 18 are rock, 24 are classical and 30 are rap music.

a) What is the ratio of rock music to rap music?

....

b) What is the ratio of rap music to classical music?

....

2 marks

18 The first four terms of a sequence are 4, 10, 20 and 34.

For each expression in the table, put a cross if it cannot be the nth term for this sequence or a tick if it could be the nth term. If you put a tick, work out the 5th term in the sequence. The first one has been done for you.

Expression	Could it be the nth term?	5th term
$3n + 1$	✗	
$2(n^2 + 1)$		
$n + 6$		
$2n^2 + 2$		

3 marks

19 A recipe for 12 cupcakes needs 100g of butter and 150g of flour. How much butter and flour are needed to make:

a) 24 cupcakes?

_____

b) 30 cupcakes?

_____

4 marks

20 **a)** Find the area of the rectangle.

4cm

16cm

_____

b) A square has the same area as the rectangle. What is the side length of the square?

_____

c) A different square has the same perimeter as the rectangle. What is the side length of this square?

_____

3 marks

TOTAL

20

21 Martin collected some data for his biology project.

He measured the width of the leaves from one of the trees in his garden.

KEY: 7 | 4 = 7.4cm

6	1 4 5 7 9
7	1 2 4 5 7 8 9
8	2 2 5 6 8
9	0 2 4

a) Show that the range of the width of the leaves is 3.3cm.

....

b) Find the mean width.

.... cm

2 marks

22 Work out the missing numbers. In each case, use the first line to help you.

a) $17 \times 15 = 255$

$17 \times$ .... $= 510$

b) $255 \div 17 = 15$

$255 \div$ .... $= 30$

c) $35 \div 2 = 17.5$

$35 \times \frac{1}{2} =$ ....

3 marks

23 Write in the correct operations in these statements.

You can use $+$, $-$, $\times$ or $\div$.

C .... C $= 0$

C .... C $= C^2$

C .... C $= 1$

C .... C $= 2C$

2 marks

24 Find the missing part of the fraction in these equivalent fractions.

a) $\frac{2}{9} = \frac{?}{45}$

b) $\frac{3}{5} = \frac{18}{?}$

c) $\frac{12}{?} = \frac{1}{12}$

3 marks

25 Find the missing angles.

a)

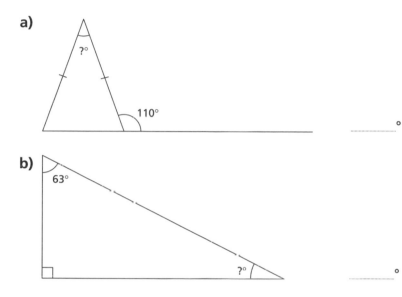

_____ °

b)

_____ °

4 marks

26 The rules show how to get from one term in the sequence to the next. Use the rule to write down the next two terms in the sequence.

a) Rule: multiply by 4

3, 12, _____, _____

b) Rule: multiply by 2 and add 1

5, 11, _____, _____

c) Rule: divide by 2 and add 9

10, 14, _____, _____

3 marks

TOTAL

17

Mixed Test-Style Questions

Calculator Allowed

1 Find the volume of the shapes.

a)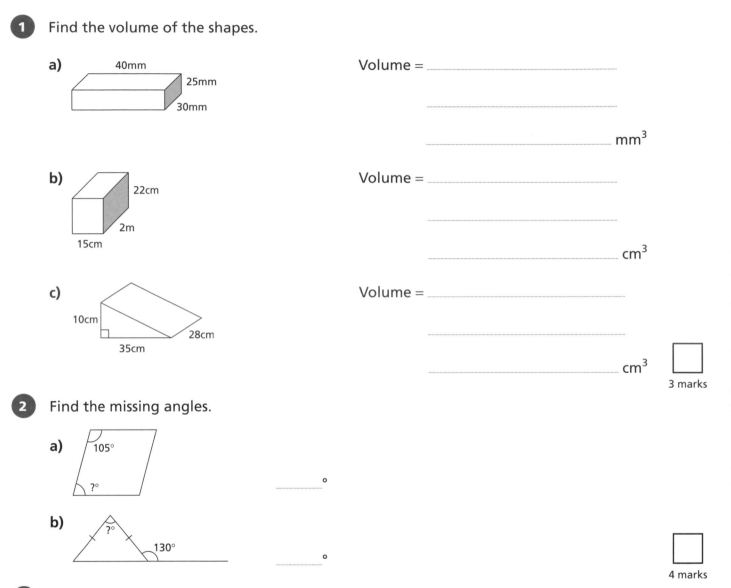

40mm

25mm

30mm

Volume = ...

....

.... mm³

b)

22cm

2m

15cm

Volume = ...

....

.... cm³

c)

10cm

28cm

35cm

Volume = ...

....

.... cm³

☐ 3 marks

2 Find the missing angles.

a)

105°

?°

.... °

b)

?°

130°

.... °

☐ 4 marks

3 The table gives information about average hours of sunshine per month in three places in Europe (A, B and C).

	Average hours of sunshine per month	Number of months	Months
A	480	7	April–October
	360	5	November–March
B	390	8	March–October
	270	4	November–February
C	420	6	May–October
	450	6	November–April

Which place has more hours of sunshine on average over the whole year? Tick the correct box and show working to justify your answer.

A ☐ B ☐ C ☐

3 marks

4 Increase £250 by 20%.

....

2 marks

5 Use Pythagoras' Theorem to work out which triangle does not have a right angle. Tick the correct box.

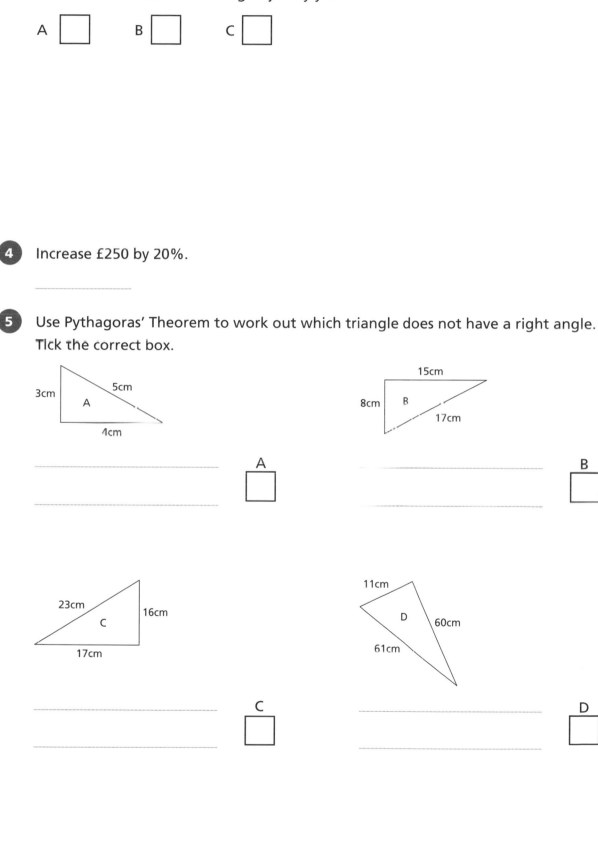

3cm 5cm A 4cm

15cm 8cm B 17cm

.... A ☐

.... B ☐

23cm 16cm C 17cm

11cm D 60cm 61cm

.... C ☐

.... D ☐

4 marks

TOTAL ☐

16

6 Mark picks cards at random from a normal deck of playing cards. What is the probability of him picking:

a) a heart? ....

b) a black card? ....

c) a King of any suit? ....

3 marks

7 Work out the size of angle y. Give your answer to 1 decimal place.

13m

5m

.... °

2 marks

8 Solve the equations.

a) $8 + 6x = 56$

b) $7x + 2 = 3x - 10$

2 marks

9 What is the interior angle sum of a heptagon?

.... °

1 mark

10 Round the following numbers to 1 significant figure.

a) i) 4567 ....

ii) 2345 ....

iii) 5.678 ....

b) Use your answers to part a) to estimate the following calculation.

$$\frac{4567 - 2345}{5.678}$$

....

c) Use your calculator to find the exact answer to part b). Give your answer to 1 decimal place.

....

6 marks

11 To make orange paint, yellow and red paint are mixed in the ratio 5 : 3.

How much yellow paint would need to be mixed with:

a) 6 litres of red paint? ....

b) 12 litres of red paint? ....

c) 21 litres of red paint? ....

3 marks

12 What is the surface area of the shape?

Surface area = ... cm^2

8cm

20cm

15cm

4 marks

13 The Robinsons are travelling to France for their summer holiday. When they leave, the exchange rate is £1 = €1.32

Mr Robinson changes £500. How many euros will he get?

....

1 mark

14 What is the height of a cylinder if its volume is 9047.79m^3 and its radius is 12m?

....

2 marks

15 Isaac has three different parcels to send in the post as below.

5cm × 10cm × 30cm   15cm × 10cm × 20cm   11cm × 9cm × 25cm

BOX 1   BOX 2   BOX 3

a) What is the minimum amount of brown wrapping paper Isaac will need for all three parcels?

....

....

b) The delivery company that Isaac is using charges by volume. Which will be the most costly parcel to deliver? Show your working.

....

7 marks

....

TOTAL

31

Mixed Test-Style Questions

16 The pie chart shows what 120 people do in their spare time.

Legend:
- ■ Books
- □ Music
- ■ Sports
- ■ Gardening
- ■ Other

Values shown: 9, 9, 36, 48, 18

a) What do you notice about the popularity of the different activities shown in the pie chart?

....

....

....

b) What angle represents sports?

....

2 marks

17 If a cuboid has a volume of 120m^3, what could the height, width and length be?

Height: ....

Width: ....

Length: ....

1 mark

18 Calculate the density of:

a) a gold bar with a mass of 350kg and a volume of 0.04m^3

....

b) an aluminium block with a mass of 900kg and a volume of 0.2m^3.

....

4 marks

196   KS3 Maths Workbook

19 **a)** Complete the table and plot the graph of $y = 4x - 3$.

x	−3	−2	−1	0	1
y					

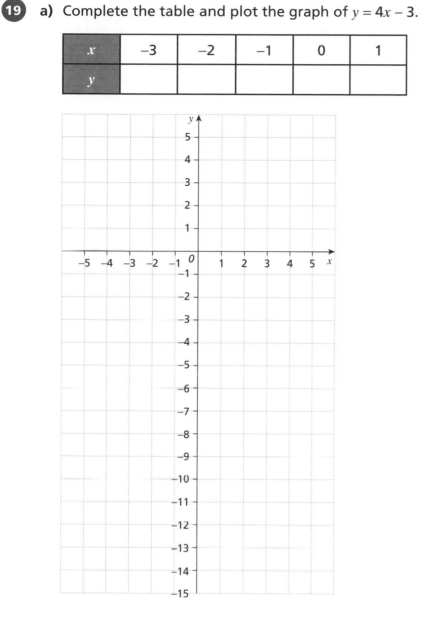

b) What is the y-intercept of the graph $y = 8x + 13$? ....

c) What is the gradient of the graph $y = 7x - 3$? ....

7 marks

20 Solve the equation $3(x + 1) = 2 + 4(2 - x)$

....

....

....

....

3 marks

TOTAL

17

21 Patrick counts the number of different coloured lunch boxes in his school dining room. From the data, he calculates the probabilities in the table.

Yellow	Black	Red	Blue	Green
0.2	0.3	0.1	0.2	

a) Fill in the missing value.

b) What is the probability of somebody's lunch box not being black?

....

c) If Patrick counts 20 red lunch boxes, how many people does he record in total?

....

5 marks

22 The graph shows the exchange rate between British pounds (£) and US dollars ($).

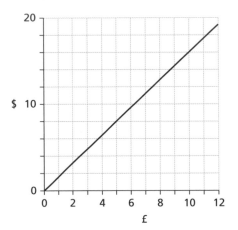

a) How many dollars would you get for:

    i) £2? ....

    ii) £10? ....

    iii) £100? ....

b) How many pounds would you get for:

    i) $10? ....

    ii) $8? ....

5 marks

23 A wire of 22m in length runs from the top of a pole to the ground as shown in the diagram. The wire makes an angle of 35° with the ground.

Calculate the height of the pole. Give your answer to a suitable degree of accuracy.

....

....

24 Lewis takes a trip in a hot air balloon. The balloon rises 600m in one hour and stays at this height for two and a half hours. The balloon then comes back to earth in half an hour.

Use the grid to draw a distance–time graph for the balloon flight.

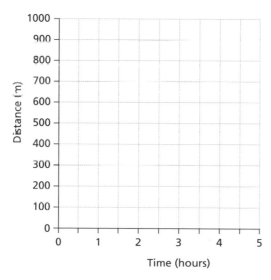

3 marks

25 During a survey, 45 people are asked a question. 10 of them choose the same answer.

On a pie chart, what angle would represent these 10 people?

....

2 marks

TOTAL

17

Mixed Test-Style Questions

26 Joan is making custard.

Her recipe is $\frac{1}{6}$ paste and the rest milk.

a) What fraction of milk does Joan need?

_____

b) If Joan makes 600ml of custard, how much milk does she use?

_____ ml

c) If Joan has 300ml of milk, how much custard paste does she need?

_____ ml

6 marks

TOTAL

6

KS3 Standard Maths Workbook Answers

Pages 148–149

Number

1. −4, −2, 0, 5, 7 **[2] [1 mark for any three correct]**
2. £1920 **[3] [2 marks for 1800 and 120; 1 mark for either 1800 or 120]**
3. 497 as 3 is less than 9 **[2] [1 mark for 3 and 9]**
4. > [1]

1. a) −4 and +5 [1]

> Remember: $[-] \times [+] = [-]$

   b) +4 and −5 [1]

2.
> The question is asking you to find the LCM of 6 and 14.

   Tram A: 6; 12; 18; 24; 30; 36; 42
   Tram B: 14; 28; 42 [1]
   LCM = 42 [1]
   9.00am + 42 minutes = 9.42am [1]

3. a) 36 **[3] [2 marks for valid attempt at division with only one numerical error; 1 mark for valid attempt at division with two numerical errors]**
   b) Halve the answer from part a). [1]
4. a) $(5 + 4) \times 7 - 3 = 60$ [1]
   b) $(2 + 3) \times (5 - 7) = -10$ [1]
5.

Operation	Inverse operation	
Add 2	Subtract 2	
Subtract 7	**Add 7**	**[1]**
Multiply by 6	**Divide by 6**	**[1]**
Divide by 4	**Multiply by 4**	**[1]**

Pages 150–151

Sequences

1. [−3, 1] [5, 9] [13, 17] **[2] [1 mark for any two correct]**
2. $n + 2$ with 3, 4, 5, 6, 7; $5n - 3$ with 2, 7, 12, 17, 22 and $4n$ with 4, 8, 12, 16, 20 **[2] [1 mark for any one correct]**

1. a) 12 **[2] [1 mark for 48]**
   b) $4n + 2$ **[2] [1 mark for 4n]**
2. a) $2n + 3$ **[2] [1 mark for 2n]**
   b) Yes **[1]**, as there is a common difference between consecutive terms **[1]**
3. a) 5, 8, 11, 14, 17 [2]
      [1 mark for any three correct]
   b) + 3 [1]
4. a) 13, 21, 34 [1]
   b) Fibonacci [1]
   c) 16, 25, 36 [1]
   d) Square numbers [1]

Pages 152–153

Perimeter and Area

1.

	Perimeter (cm)	Area (cm^2)
Square A	24	36
Rectangle B	18	18
Square C	12	9

   a) Table completely correct **[3]**
      [2 marks for any four correct; 1 mark for any two correct]
   b) 18cm^2 [1]
2. 4cm [1]

1. Tomiwa is wrong **[1]** – he works out the circumference, not the area. **[1]**

2. 9200m² **[3]**

[2 marks for 6000 and 3200 *or* 10 000 and 800; 1 mark for either 6000 *or* 3200 *or* 10 000 *or* 800]

> Remember that this can be split into a rectangle and trapezium or a rectangle minus a triangle.

3. **a)** 0.24m²   **[2]**

[1 mark for 60 × 40 or 0.6 × 0.4]

b) 2 tins   **[2]**

[1 mark for 0.24 ÷ 0.15]

Pages 154–155
Statistics and Data

▶

1. **a)** 5   **[1]**

b) 5   **[1]**

> Remember to write the numbers in order first.

c) 4.4 **[2]** **[1 mark for 31]**

2.

Time	Tally	Frequency
$10 \leqslant t < 20$	⅃⅃⅃⅃ ⅃⅃⅃⅃ I	11
$20 \leqslant t < 25$	⅃⅃⅃⅃ IIII	9
$25 \leqslant t < 30$	⅃⅃⅃⅃ ⅃⅃⅃⅃ ⅃⅃⅃⅃ II	17
$30 \leqslant t < 50$	⅃⅃⅃⅃ IIII	9

Fully correct **[3]**

[2 marks for correct tallies and one mistake in frequency; 1 mark for three out of four correct tallies]

▶

1. **a)** Both of them   **[1]**

b) Bimodal   **[1]**

2. Yes, because the new median is 6.5.   **[1]**

3. **a)** 6, 6, 11 or 1, 6, 6 **[2]**

[1 mark for a set of numbers which contains at least two 6s]

b) 10, 11, 12   **[1]**

4. **a)** Table completely correct **[2]**

[1 mark for four out of six correct]

	Rock	Pop	Classical	Total
Male	18	15	7	40
Female	7	27	6	40
Total	25	42	13	80

b) Any correct statement, for example:
More men preferred **rock** *or* **classical** than women.   **[1]**
Rock was **more** popular than **classical**.   **[1]**
or
Pop was **more** popular than **both rock** and **classical**.   **[1]**

Pages 156–157
Decimals

▶

1. 1.37 and 0.63; 1.44 and 0.56; 1.49 and 0.51
[2] [1 mark for one correct]

2. × 1000 **[1]**; ÷ 100 **[1]**

▶

1. 4 and 0.5; 8 and 0.25; 10 and 0.2; 100 and 0.02 **[2] [1 mark for one correct]**

2. **a)** 17.001, 17.01, 17.9007, 17.91   **[1]**

b) 3 046 491   **[1]**

c) 304.6491   **[1]**

3. True **[1]**, they are equivalent fractions. **[1]**

4. Pack of 5 with 70 and 65 *or* 0.70 and 0.65 **[2] [Same addition for 1 mark]**

5. 263   **[1]**

> Remember 5.26 ÷ 0.02 = 526 ÷ 2

Pages 158–159
Algebra

▶

1. **a)** $2x + 2x + 1 - 3$ **[1]**

b) Because it simplifies to $4x - 2$, not $4x + 2$.   **[1]**

2. **a)** a^3 **[1]**

b) $5b$ **[1]**

c) c^2d^2 **[1]**

3. He has only multiplied the x by 2 and he should have multiplied the 5 by 2 as well.   **[1]**

4. x **[1]**

1. 9, 15 and 7 **[2] [1 mark for any two correct]**

   Remember BIDMAS.

2. a) −12 **[1]**
   b) Yes **[1]**, as the first expression has been factorised to make the second. **[1]**

3. a) $3y + 6$ **[1]**
   b) $6v − 6$ **[2]** or $2v + 2 + 4v − 8$
   [1 mark for any three correct terms]

4. a) $5(2x + 1)$ **[1]**
   b) $x(2x + 1)$ **[1]**

Pages 160–161

3D Shapes: Volume and Surface Area

1.

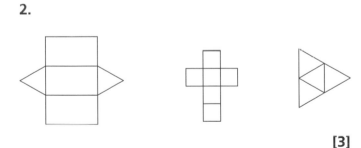

Shape	Name	Vertices	Edges	Faces
	Triangular prism	6	9	5
	Cube	8	12	6
	Tetrahedron or triangular-based pyramid	4	6	4

[6] [$\frac{1}{2}$ mark for each correct answer]

2.

[3]

3. Cuboid A:
   Surface area: $2(6 \times 3) + 2(3 \times 11) +$
   $2(6 \times 11) = 234$cm^2 **[2]**
   Volume: $6 \times 3 \times 11 = 198$cm^3 **[2]**
   Cuboid B:
   Surface area: $2(14.5 \times 2) + 2(2 \times 7) +$
   $2(14.5 \times 7) = 289$cm^2 **[2]**
   Volume: $14.5 \times 2 \times 7 = 203$cm^3 **[2]**
   [2 marks for each correct answer; 1 mark for method]

1. Cylinder A:
   Surface area: $2(7^2 \times \pi) + (14 \times \pi) \times 6$
   $= 571.8$cm^2 **[2]**
   Volume: $7^2 \times \pi \times 6 = 923.6$cm^3 **[2]**
   Cylinder B:
   Surface area: $2(2^2 \times \pi) + (4 \times \pi) \times 25$
   $= 339.3$cm^2 **[2]**
   Volume: $2^2 \times \pi \times 25 = 314.2$cm^3 **[2]**
   [2 marks for each correct answer; 1 mark for method]

2. $8 \times 4 \times ? = 64$ **[1]**
   $64 \div 32 = 2$cm **[1]**

3. $1357.2 \div 12 = 113.1$
   $113.1 \div \pi = 36$ **[1]**
   $\sqrt{36} = 6$m **[1]**

4. Container A as the volume is greater.
   Container A: $6^2 \times \pi \times 15 = 1696.5$cm^3
   Container B: $10 \times 12 \times 12 = 1440$cm^3 **[4]**
   [1 mark for each method and each answer]

Pages 162–163

Interpreting Data

1.

Category	Frequency	Angle
Polo Mints	8	$\frac{8}{60} \times 360 = \mathbf{48°}$
Wine Gums	12	$\frac{12}{60} \times 360 = \mathbf{72°}$
Gummi Bears	17	$\frac{17}{60} \times 360 = \mathbf{102°}$
Others	23	$\frac{23}{60} \times 360 = \mathbf{138°}$

[3 marks if all correct; 2 marks if three rows correct; 1 mark if two rows correct]

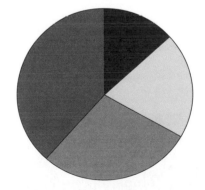

■ Polo Mints □ Wine Gums ■ Gummi Bears ■ Others

[2 marks if all sectors correct to within 2°; 1 mark if two correct]

> If one angle has to be rounded down to fit into a pie chart, round the largest one so it is the least proportional error.

2. a) Tuesday, $9 + 4 = 13$ [1]
    b) Gareth $= 12 + 4 + 15 + 11 + 8 + 34 + 30$
    $\qquad = 114$
    Rachel $= 4 + 9 + 12 + 16 + 14 + 24 + 27$
    $\qquad = 106$ [1]
    Gareth sold the most. [1]

1. E.g. ice-cream sales and temperature; fuel use and engine size; drink sales and sporting activities [2]
2. Question must be specific, using a time reference; answer boxes cannot overlap and must cover all options. [3]
    [1 mark for each correct part]
3. Sales at the weekend were significantly increased for both. [1]
    Both had their least amount of sales on Tuesday. [1]
    Rachel outsold Gareth on three days – Tuesday, Thursday and Friday. [1]
    Gareth outsold Rachel on four days – Monday, Wednesday, Saturday and Sunday. [1]
4. No **[1]**, Jordan spends one hour more on sporting activities. **[1]**

Pages 164–165

Fractions

1. a) 9 [1]
    b) 30 [1]
    c) 12 [1]
2. a)
    [1]
    b) $1\frac{1}{5}$ **[2] [1 mark for 1 and 1 mark for $\frac{1}{5}$]**
3.
    $\frac{3}{4}$ [2]

[1 mark for both shapes correctly shaded and 1 mark for $\frac{3}{4}$]

4. Two-and-a-half pizzas [2]
    [1 mark for stating two whole ones; 1 mark for the half]

1. a) $\frac{11}{14}$ [1]
    b) $\frac{25}{36}$ [1]
    c) $\frac{4}{5}$ [1]
    d) $\frac{9}{14}$ [1]
    e) $\frac{1}{3}$ [1]
    f) $\frac{33}{70}$ [1]
2. a) $\frac{7}{10}$ [1]
    b) $\frac{5}{24}$ [1]
    c) $\frac{15}{56}$ [1]
    d) $\frac{2}{7}$ [1]
    e) $\frac{5}{48}$ [1]
    f) $\frac{16}{225}$ [1]
    g) $\frac{2}{15}$ [1]
    h) $1\frac{3}{5}$ [1]
    i) $1\frac{13}{14}$ [1]
    j) $2\frac{1}{27}$ [1]
3. $2\frac{1}{3} + 2\frac{1}{7} = 4 + \frac{7}{21} + \frac{3}{21}$ [1]
    $4\frac{10}{21}$ [1]
    $4\frac{3}{4} - 4\frac{10}{21} = \frac{23}{84}$ [1]
4. $5\frac{3}{10} - 1\frac{1}{2}$ [1]
    $\frac{53}{10} - \frac{3}{2}$ [1]
    $\frac{19}{5} = 3\frac{4}{5}$pts [1]

Pages 166–167

Coordinates and Graphs

1. a)

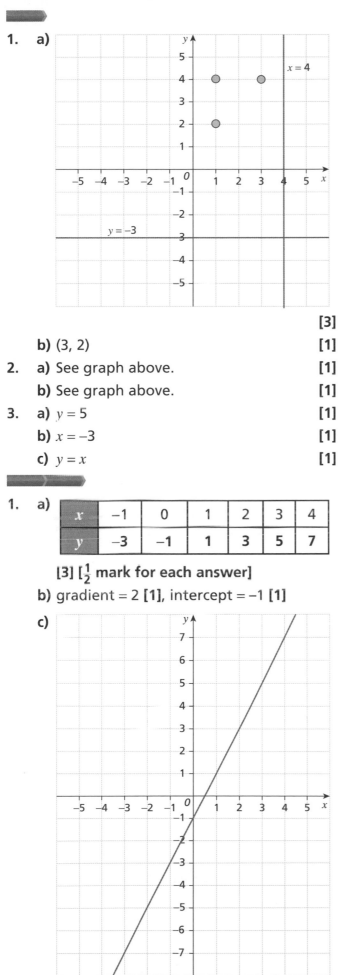

[3]

   b) (3, 2) [1]
2. a) See graph above. [1]
   b) See graph above. [1]
3. a) $y = 5$ [1]
   b) $x = -3$ [1]
   c) $y = x$ [1]

1. a)

x	−1	0	1	2	3	4
y	−3	−1	1	3	5	7

   [3] [$\frac{1}{2}$ mark for each answer]
   b) gradient = 2 [1], intercept = −1 [1]
   c)

[1 mark if graph passes through one correct coordinate]

2. a) $y = 3x +$ (any number) [1]
   b)

x	y
1	5
2	8
3	11
0	2
−1	−1
−2	−4
−3	−7

[1 mark for any pair of coordinates]

Pages 168–169

Angles

1. a) acute – 80° [2]
   b) obtuse – 155° [2]
   c) acute – 25° [2]
   d) reflex – 320° [2]
   [1 mark for each angle; 1 mark for each label]
2. 23° and 157° [1]; 119° and 61° [1]; 99° and 81° [1]
3. [1 mark for construction lines on the angle lines; 1 mark for construction cross; 1 mark for bisecting line]

1.

Number of sides	Name of shape	Sum of interior angles
3	Triangle	180°
5	Pentagon	540°
6	Hexagon	720°
9	Nonagon	1260°

[4]

[1 mark for each correct row]

2. a) 180° − 135° = 45° [1]
      180° − 45° − 45° = 90° [1]
   b) 180° − 150° = 30
      360° − 30° − 30° = 300° [1]
      300° ÷ 2 = 150° [1]

c) $180° - 73°$ **[1]** $= 107°$ **[1]**

d) $180° - 54°$ **[1]** $= 126°$ **[1]**

> Corresponding angles, then angles on a straight line add up to 180°.

Pages 170–171

Probability

1.

Impossible	Unlikely	Even chance	Likely	Certain
A	D	E	C	B

[5]

2. a) $\frac{7}{20}$ **[1]**

b) $\frac{13}{20}$ **[1]**

3. a)

Ham sandwich	Cheese sandwich	Chicken wrap	Other
0.3	0.25	0.4	**0.05**

[1]

b) $1 - 0.3 = 0.7$ **[1]**

1. a) $\frac{3}{12} = \frac{1}{4}$ **[1]**

b) $\frac{1}{11}$ **[1]**

2. $1 - (0.4 + 0.35)$ **[1]** $= 0.25$ **[1]**

3. Possible answer: 4 winning segments give a probability of breaking even **[1]**, so no more than 3 **[1]**, and possibly fewer, should raise more money. **[1]**

Pages 172–173

Fractions, Decimals and Percentages

1.

Fraction	Decimal	Percentage
$\frac{3}{10}$	0.3	30
$\frac{45}{100} = \frac{9}{20}$	**0.45**	45
$\frac{72}{100} = \frac{18}{25}$	0.72	**72**
$\frac{9}{100}$	0.09	9

[4]

[3 marks for any six correct; 2 marks for any four correct; 1 mark for any two correct]

2. 5.25 **[1]**

3. 50% of £720 = £360 **[1]**; 25% = £180 **[1]**

4. $1500 ÷ 3 × 2$ **[1]** $= 1000$ml **[1]**

5. 10% = £43; £430 − £43 **[1]** = £387 **[1]**

6. 10% = £43; 20% = £86 **[1]**; £430 + £86 = £516 **[1]**

7. a) The increase in value is £4000. **[1]**

   The fraction is $\frac{4000}{80\,000}$ **[1]**

   b) $\frac{1}{20}$ **[1]**

1. a) 10% = 5000 **[1]**

   $50\,000 + 5000 = 55\,000$ bacteria **[1]**

   b) 10% = 5500 **[1]**

   $55\,000 + 5500 = 60\,500$ bacteria **[1]**

2. Anna: $38 ÷ 50 × 100 = 76\%$

   Ben: $32 ÷ 40 × 100 = 80\%$ **[1]**

   Ben scored the highest percentage. **[1]**

3. a) $0.04 × 500 = 20$ **[1]**

   $500 + 20 = £520$ **[1]**

   b) $20 × 3$ yrs = £60 **[1]**

   $500 + 60 = £560$ **[1]**

4. 80% of 45 minutes

   $80 ÷ 100 × 45$ **[1]**

   = 36 minutes **[1]**

5.

> To compare the quantities, the units must be the same.

3kg $= 3000$g

$240 ÷ 3000 × 100$ **[1]**

$= 8\%$ **[1]**

6. $30\% = \frac{30}{100} = \frac{3}{10}$ and $\frac{3}{10} \neq \frac{1}{3}$

   or $\frac{1}{3} = 0.333333 = 33.33\%$ **[1]**

Pages 174–175

Equations

1. a) 3 **[1]**

   b) 8 **[1]**

   c) 7 **[1]**

2. $(− 6)$ $2a = 4$ **[1]**

   $(÷ 2)$ $a = 2$ **[1]**

3. A and D; B and F; C and E **[3]**

4. a) $2y = 12$ **[1]**

   $y = 6$ **[1]**

b) $3a = 27$ [1]

$\quad a = 9$ [1]

c) $4p = 12$ [1]

$\quad p = 3$ [1]

5. a) $6x + 25 = 235$ [1]

b) $6x + 25 = 235$

$\quad (-25)\ 6x = 210$ [1]

$\quad (\div 6)\quad x = 35$

David worked for 35 hours. [1]

1. $7(n - 3) = 4n + 3$ [1]

$\quad 7n - 21 = 4n + 3$

$\quad 3n - 21 = 3$ [1]

$\quad\quad 3n = 24$

$\quad\quad n = 8$

I am thinking of the number 8. [1]

2. a) $7y - 4 = 2y + 31$

$\quad 5y - 4 = 31$ [1]

$\quad 5y = 35$

$\quad y = 7$ [1]

b) $8a + 5 = 3a + 20$

$\quad 5a + 5 = 20$ [1]

$\quad 5a = 15$

$\quad a = 3$ [1]

c) $7p + 4 = 4p - 2$

$\quad 3p + 4 = -2$ [1]

$\quad 3p = -6$

$\quad p = -2$ [1]

3. $(+ 6)\ \dfrac{3n}{4} = 9$ [1]

$\quad (\times 4)\ 3n = 36$

$\quad (\div 3)\ n = 12$ [1]

4. a) $4(2x - 1) = 36$

Expand the brackets first.

$\quad 8x - 4 = 36$ [1]

$\quad 8x = 40$

$\quad x = 5$ [1]

or

$\quad 2x - 1 = \dfrac{36}{4} = 9$

$\quad 2x = 10$

$\quad x = 5$

b) $3x + 8 = 4 - 5x$

$\quad 8x + 8 = 4$ [1]

$\quad 8x = -4$

$\quad x = -0.5$ [1]

c) $\dfrac{3x - 6}{4} = 3$

$\quad 3x - 6 = 12$ [1]

$\quad 3x = 18$

$\quad x = 6$ [1]

d) $\dfrac{x}{2} + 3 = \dfrac{x}{4} + 5$

Remember to multiply every term.

$(\times 2)\ x + 6 = \dfrac{2x}{4} + 10$ [1]

$(\times 4)\ 4x + 24 = 2x + 40$ [1]

$\quad 2x = 16$

$\quad x = 8$ [1]

5. $3x - 2 + 3x - 2 + 2x + 2x = 56\text{cm}$

[1 mark for writing a correct expression]

$10x - 4 = 56\text{cm}$

[1 mark for simplifying the expression]

$10x = 60\text{cm}$

$\quad x = 6\text{cm}$ [3]

Pages 176–177

Symmetry and Enlargement

1. a)   **b)**   **c)**

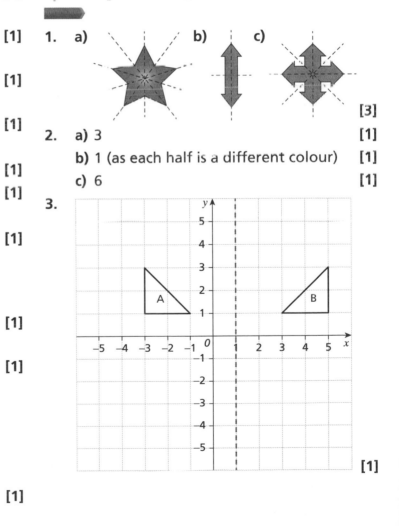

[3]

2. a) 3 [1]

b) 1 (as each half is a different colour) [1]

c) 6 [1]

3.

[1]

1.

a) See above [1]

b) See above [1]

c) $180°$ about the point (0, 0) [1]

d) See above [2]

2. Scale factor is 3. [1]

Wing span $= 21 \div 3 = 7$cm [1]

Pages 178–179

Ratio and Proportion

1. 9 : 6 *or* 3 : 2 [1]

2. a) 16 : 12 *or* 4 : 3 [1]

b) 3 : 4 [1]

3. a) 1 : 2 [1]

b) 3 : 4 [1]

c) 40 minutes : 60 minutes = 2 : 3 [1]

d) 15mm : 40mm = 3 : 8 [1]

4. a) 28 : 40 [1]

b) 7 : 10 [1]

1. $700 \div 7 = 100$ [1]

3 parts $= 100 \times 3 = 300$g

4 parts $= 100 \times 4 = 400$g

$= 300$g : 400g [1]

2. a) £16 ÷ 2 = £8 [1]

The larger share = £8 × 5 = £40 [1]

b) Total amount to be shared

$= £16 + £40$ **[1]** $= £56$ **[1]**

3. a) 100g × 2 = 200g flour [1]

b) 2 eggs × 2 = 4 eggs [1]

c) 100g ÷ 12 × 30 = 250g sugar [1]

or

100g for 12 cakes

200g for 24 cakes

50g for 6 cakes

250g for 30 cakes

4. 1 pencil costs £1.60 ÷ 8 = 20p [1]

3 pencils cost 20p × 3 = 60p [1]

5. 120cm – 45cm = 75cm [1]

Ratio is 45 : 75 [1]

= 3 : 5 [1]

Pages 180–181

Real-Life Graphs and Rates

1. a) 80km [1]

b) 32km (±1km) [1]

c) 160km [1]

2. a)

Distance travelled (miles)	0	5	10	15	20	25
Petrol used (litres)	0	2	4	6	8	10

[2] [1 mark for three correct]

b) 1 mile uses 2 ÷ 5 = 0.4 litres of petrol [1]

4 miles use 0.4 × 4 = 1.6 litres of petrol

(also accept 1600ml/160cl) [1]

1. a) 50km [1]

b) 2.5 hours [1]

c) 1 hour [1]

d) Speed = total distance ÷ total time

$= 50 \div 0.5$ **[1]** $= 100$km/h **[1]**

2. 30 × 4 **[1]** = 120 miles **[1]**

3. 200 × 1.22 **[1]** = €244 **[1]**

4. 200 ÷ 1.22 **[1]** = £163.93 **[1]**

5. Volume = 15 000 ÷ 700 **[1]** = 21.43m^3 **[1]**

Pages 182–183

Right-Angled Triangles

1. a) $8^2 + 15^2 = p^2$

$64 + 225 = 289$ [1]

$p = \sqrt{289} = 17$cm [1]

b) $24^2 + 7^2 = q^2$

$576 + 49 = 625$ [1]

$q = \sqrt{625} = 25$cm [1]

2. a) $c^2 + 14^2 = 50^2$

$c^2 = 2500 - 196$ [1]

$c = \sqrt{2304} = 48$cm [1]

b) $12^2 + d^2 = 19^2$

$d^2 = 361 - 144$ [1]

$d = \sqrt{217} = 14.7\text{cm}$ [1]

1. $7^2 + 24^2 = c^2$

$49 + 576 = 625$

$c = \sqrt{625} = 25\text{cm}$ [1]

> Now use the second triangle.

$y^2 + 20^2 = 25^2$ [1]

$y^2 = 625 - 400$

$y = \sqrt{225} = 15\text{cm}$ [1]

2. **a)** 0.87 [1]

b) 16.58 [1]

c) 21.42 [1]

3. **a)** $\tan 67^\circ = 15 \div x$

$x = 15 \div \tan 67^\circ$ [1]

$x = 6.4\text{cm}$ [1]

b) $\sin 12^\circ = x \div 22$

$x = \sin 12^\circ \times 22$ [1]

$x = 4.6\text{cm}$ [1]

4. **a)** $\cos x^\circ = 39 \div 57$

$x = \cos^{-1}(39 \div 57)$ [1]

$x = 46.8^\circ$ [1]

b) $\sin x^\circ = 8 \div 12$

$x = \sin^{-1}(8 \div 12)$ [1]

$x = 41.8^\circ$ [1]

5. **a)** $\cos 44^\circ = h \div 5$ [1]

$\cos 44^\circ \times 5 = h = 3.6\text{m}$ [1]

b) $\sin 44^\circ = \text{base} \div 5$ [1]

$\sin 44^\circ \times 5 = \text{base} = 3.5\text{m}$ [1]

Pages 184–200

Mixed Test-Style Questions

No Calculator Allowed

1. **a) i)** 28 [1]

ii) The median is between 25 and 27,

so is 26. [1]

b) The mean [1] or range [1]

2. **a)** $\frac{29}{35}$ [1]

b) $\frac{35}{72}$ [1]

c) $\frac{33}{70}$ [1]

3. £840 ÷ 12 = £70 [1]

3 parts = 3 × £70 = £210

4 parts = 4 × £70 = £280

5 parts = 5 × £70 = £350 [1]

£210 : £280 : £350 [1]

4. $x = 3$ **[2] [1 mark for 25 and 100]**

5. **a) [3]** and **b) [2]**

c) (3, 1) *or* (0, 5) *or* (5, 2) [1]

6. **a)** $\frac{4}{14} = \frac{2}{7}$ [1]

b) Special J as there is a $\frac{1}{14}$ chance of

him selecting it. [1]

c) $1 - \frac{2}{14}$ **[1]** $= \frac{12}{14} = \frac{6}{7}$ **[1]**

7. 1% = 200 ÷ 100 = £2

4% = £2 × 4 = £8 [1]

After 3 years: £200 + (3 × £8) = £224 [1]

8. **a)** $\frac{3}{4}$ **[1]** 75% **[1]**

b) $\frac{3}{10}$ **[1]** 30% **[1]**

9. **a)** $a + 2b$ **[2] [1 mark for 2a + 4b]**

b) 8 **[2] [1 mark for 32]**

10.

[1 mark for correct position on *x*-axis;

1 mark for correct position on *y*-axis]

11. a) £9.75 [1]
    b) £10 [1]
    c) 35% [1]
    d) £20 [1]

12.

    [2 marks if fully correct; 1 mark for any two correct]

13. 4^2 as $4^2 = 16$ but $2^3 = 8$ [1]

14. $S = 10 \div 0.5$ [1] $= 20$mph [1]

15.

    a) Reflection [1]
    b) Enlargement [2]
    c) Rotation [2]

16. $4x + 12 = 40$ [1]
        $4x = 28$ [1]
        $x = 7$ [1]

17. a) 18 : 30 or 3 : 5 [1]
    b) 30 : 24 or 5 : 4 [1]

18.

Expression	Could it be the nth term?	5th term
$3n + 1$	✗	
$2(n^2 + 1)$	✔	52
$n + 6$	✗	
$2n^2 + 2$	✔	52

[1 mark for each correct row]

19. a) 24 cupcakes – butter: 100g $\times 2 = 200$g [1]
        flour: 150g $\times 2 = 300$g [1]
    b) 30 cupcakes – butter: 200g $+ 50$g
        $= 250$g [1], flour: 300g $+ 75$g $= 375$g [1]

20. a) 64cm^2 [1]
    b) 8cm [1]
    c) 10cm [1]

21. a) $9.4 - 6.1 = 3.3$ [1]
    b) 7.76 [1]

22. a) 30 [1]
    b) 8.5 [1]
    c) 17.5 [1]

23. $C - C = 0$
    $C \times C = C^2$
    $C \div C = 1$
    $C + C = 2C$
    [2] [1 mark for any two correct]

24. a) 10 [1]
    b) 30 [1]
    c) 144 [1]

25. a) $180 - 110 = 70$ [1]
        $180 - 70 - 70 = 40°$ [1]
    b) $180 - 90 = 90$ [1]
        $90 - 63 = 27°$ [1]

26. a) 48 and 192 [1]
    b) 23 and 47 [1]
    c) 16 and 17 [1]

Mixed Test-Style Questions
Calculator Allowed

1. a) $30\,000$mm^3 [1]
   b) $15 \times 200 \times 22 = 66\,000$cm^3 [1]
   c) 4900cm^3 [1]

2. a) Opposite angles in a parallelogram are equal.

    $360 - 105 - 105 = 150$ [1]

    $150 \div 2 = 75°$ [1]

    b) $180 - 130 = 50$ [1]

    $180 - 50 - 50 = 80°$ [1]

3. C plus 5220 is greater than 5160 and 4200 [3]

    or C plus 435 is greater than 430 and 350. [3]

    [2 marks for any two of 5220, 5160, 4200, 435, 430 and 350; 1 mark for any one of 5220, 5160, 4200, 435, 430 and 350]

4. $10\% = £25$, $20\% = £50$ [1]

    $£250 + £50 = £300$ [1]

5. A $3^2 + 4^2 = 5^2$

    $9 + 16 = 25$

    Yes [1]

    B $8^2 + 15^2 = 17^2$

    $64 + 225 = 289$

    Yes [1]

    C $17^2 + 16^2 = 23^2$

    $289 + 256 \neq 529$

    No [1]

    D $60^2 + 11^2 = 61^2$

    $3600 + 121 = 3721$

    Yes [1]

6. a) $\frac{1}{4}$ [1]

    b) $\frac{1}{2}$ [1]

    c) $\frac{4}{52} = \frac{1}{13}$ [1]

7. $\sin y = 5 \div 13$ [1]

    $y = \sin^{-1}(5 \div 13) = 22.6°$ [1]

8. a) $6x = 48$, $x = 8$ [1]

    b) $4x = -12$, $x = -3$ [1]

9. $900°$ [1]

10. a) i) 5000 [1]

    ii) 2000 [1]

    iii) 6 [1]

    b) 500 [1]

    c) 391.3 **[1 mark for 391.334 977 1]** [2]

11. a) 10 litres [1]

    b) 20 litres [1]

    c) 35 litres [1]

12. $2(8 \times 20) + 2(8 \times 15) + 2(20 \times 15) = 1160\text{cm}^2$

    [4] [1 mark for each part]

13. $500 \times 1.32 = €660$ [1]

14. $9047.79 \div \pi \div 144$ **[1]** $= 20\text{m}$ **[1]**

15. a) Box 1 = $(10 \times 5 + 10 \times 30 + 5 \times 30) \times 2 = 1000\text{cm}^2$ [1]

    Box 2 = $(15 \times 10 + 15 \times 20 + 10 \times 20) \times 2 = 1300\text{cm}^2$ [1]

    Box 3 = $(11 \times 9 + 9 \times 25 + 11 \times 25) \times 2 = 1198\text{cm}^2$ [1]

    Total = 3498cm^2 [1]

    b) Box 1 = $10 \times 5 \times 30 = 1500\text{cm}^3$

    Box 2 = $15 \times 10 \times 20 = 3000\text{cm}^3$

    Box 3 = $11 \times 9 \times 25 = 2475\text{cm}^3$

    Box 2 will cost the most.

    [2 marks for working; 1 mark for correctly stating Box 2]

16. a) Sports was the most popular choice of pastime. Gardening and other had the lowest popularity (9 each).

    [1 mark for any relevant answer]

    b) $\frac{48}{120} \times 360 = 144°$ [1]

17. Example: $2\text{m} \times 6\text{m} \times 10\text{m}$ (three numbers multiplied to make 120) [1]

18. a) Density = mass ÷ volume = $350 \div 0.04$ [1]

    $= 8750\text{kg/m}^3$ [1]

    b) Density = $900 \div 0.2$ **[1]** $= 4500\text{kg/m}^3$ **[1]**

19. a)

x	−3	−2	−1	0	1
y	−15	−11	−7	−3	1

    [3 marks if fully correct; 2 marks for four correct answers; 1 mark for two correct answers]

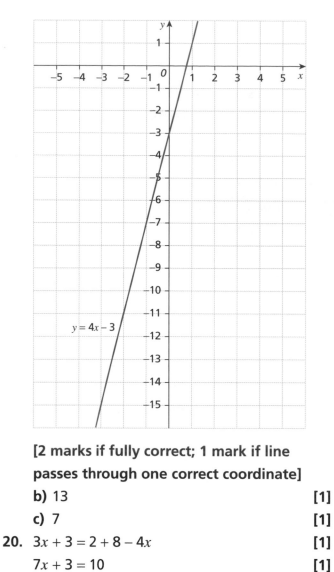

y = 4x − 3

[2 marks if fully correct; 1 mark if line passes through one correct coordinate]

b) 13 [1]

c) 7 [1]

20. $3x + 3 = 2 + 8 − 4x$ [1]

$7x + 3 = 10$ [1]

$7x = 7$

$x = 1$ [1]

21. a) 0.2 [1]

b) 1 − 0.3 [1] = 0.7 [1]

c) 20 ÷ 0.1 [1] = 200 [1]

22. a) i) $3 [1]

ii) $16 [1]

iii) $160 [1]

b) i) £6.20 (accept £6.10–£6.30) [1]

ii) £5 [1]

23. $\sin 35° = h ÷ 22$ [1]

$\sin 35° × 22 = h = 12.6m$ [1]

24.

[1 mark for each part of the graph]

25. $\frac{10}{45} × 360$ [1] = 80° [1]

26. a) $1 − \frac{1}{6}$ [1] = $\frac{5}{6}$ [1]

b) 600ml ÷ 6 = 100ml [1]

100ml × 5 = 500ml [1]

c) 300ml ÷ 5 [1] = 60ml

60ml of paste [1]

This page has been left blank

Revision Tips

Rethink Revision

Have you ever taken part in a quiz and thought '*I know this*!', but no matter how hard you scrabbled around in your brain you just couldn't come up with the answer?

It's very frustrating when this happens, but in a fun situation it doesn't really matter. However, in tests and assessments, it is essential that you can recall the relevant information when you need to.

Most students think that revision is about making sure you **know** *stuff*, but it is also about being confident that you can **retain** that *stuff* over time and **recall** it when needed.

Revision that Really Works

Experts have found that there are two techniques that help with *all* of these things and consistently produce better results in tests and exams compared to other revision techniques.

Applying these techniques to your KS3 revision will ensure you get better results in tests and assessments and will have all the relevant knowledge at your fingertips when you start studying for your GCSEs.

It really isn't rocket science either – you simply need to:

- **test yourself** on each topic as many times as possible
- **leave a gap** between the test sessions.

It is most effective if you leave a good period of time between the test sessions, e.g. between a week and a month. The idea is that just as you start to forget the information, you force yourself to recall it again, keeping it fresh in your mind.

Three Essential Revision Tips

1 **Use Your Time Wisely**
- Allow yourself plenty of time
- Try to start revising six months before tests and assessments – it's more effective and less stressful
- Your revision time is precious so use it wisely – using the techniques described on this page will ensure you revise effectively and efficiently and get the best results
- Don't waste time re-reading the same information over and over again – it's time-consuming and not effective!

2 **Make a Plan**
- Identify all the topics you need to revise (this Complete Revision & Practice book will help you)
- Plan at least five sessions for each topic
- A one-hour session should be ample to test yourself on the key ideas for a topic
- Spread out the practice sessions for each topic – the optimum time to leave between each session is about one month but, if this isn't possible, just make the gaps as big as realistically possible.

3 **Test Yourself**
- Methods for testing yourself include: quizzes, practice questions, flashcards, past-papers, explaining a topic to someone else, etc.
- This Complete Revision & Practice book gives you seven practice test opportunities per topic
- Don't worry if you get an answer wrong – provided you check what the right answer is, you are more likely to get the same or similar questions right in future!

Visit our website to download your free flashcards, for more information about the benefits of these revision techniques and for further guidance on how to plan ahead and make them work for you.

www.collins.co.uk/collinsks3revision

This page has been left blank

Acknowledgements

The authors and publisher are grateful to the copyright holders for permission to use quoted materials and images.

All images are © Shutterstock.com

Every effort has been made to trace copyright holders and obtain their permission for the use of copyright material. The authors and publisher will gladly receive information enabling them to rectify any error or omission in subsequent editions. All facts are correct at time of going to press.

Published by Collins
An imprint of HarperCollinsPublishers
1 London Bridge Street
London SE1 9GF

© HarperCollinsPublishers Limited 2020

ISBN 9780007562770

First published 2014

This edition published 2020

10

British Library Cataloguing in Publication Data.

A CIP record of this book is available from the British Library.

Publishing Manager: Rebecca Skinner
Authors: Samya Abdullah, Rebecca Evans and Gillian Spragg
Project Management: Daniel Dyer, Richard Toms, Cambridge Publishing Management Limited and Philippa Boxer
Cover Design: Sarah Duxbury and Kevin Robbins
Inside Concept Design: Sarah Duxbury and Paul Oates
Text Design and Layout: Jouve India Private Limited
Production: Robert Smith
Printed and bound in Great Britain by Bell and Bain Ltd, Glasgow

MIX
Paper from
responsible source
FSC www.fsc.org
FSC™ C007454

This book is produced from independently certified FSC™ paper to ensure responsible forest management.

For more information visit:
www.harpercollins.co.uk/green

5 EASY WAYS TO ORDER

1. Available from collins.co.uk
2. Fax your order to 01484 665736
3. Phone us on 01484 668148
4. Email us at education@harpercollins.co.uk
5. Post your order to: Collins Education, FREEPOST RTKB-SGZT-ZYJL, Honley HD9 6QZ